C000150406

THE
FRESHWATER FISHES
OF THE BRITISH ISLES

BY

C. TATE REGAN, M.A.

ASSISTANT IN THE ZOOLOGICAL DEPARTMENT OF THE
BRITISH MUSEUM (NATURAL HISTORY)

WITH TWENTY-SEVEN FIGURES AND THIRTY-SEVEN PLATES
BY THE AUTHOR

METHUEN & CO. LTD.
36 ESSEX STREET W.C.
LONDON

PREFACE

FISHES captured in the fresh waters of the British Isles are not for the most part of great commercial importance. The Salmonoids are a notable exception to this statement, for the Salmon and Trout are valued food-fishes, and the Windermere Char and the Lough Neagh Pollan are netted in numbers for the market. Smelt and Shad are also the objects of regular fisheries, and on the Severn Lampreys are still captured and potted, whilst the Eel fisheries are an important industry except in Scotland, where this fish is not appreciated. But the so-called coarse fishes, Pike, Perch, Roach, Bream, etc., are seldom exposed for sale in this country; they are not inedible; some, indeed, are very well flavoured, and are valued as food-fishes on the Continent. In former days in England, Pike, Carp, Tench, and other species were esteemed as luxuries, and every large country establishment had a pond wherein fish were kept and fattened for the table. But in those days transport was a very different affair, and so long as fresh sea-fish can be obtained at a low price in inland markets, the coarse fishes of our lakes and rivers are not likely to rise in popular estimation.

However, there are few species that do not interest the disciple of Walton, who captures the larger fish for sport and the smaller ones for bait, and it is to a large extent for the angler that this book has been written, although it is not a book about angling. It is an attempt to present in popular form the distinctive characters of our fresh-water fishes, and to give a reliable account of the main features of their life-history.

I believe that it will be found that this work is more complete than any of its kind that has yet appeared, for the different forms of Char and Whitefish receive detailed treatment, and natural hybrids are also included. In these groups I have been able to examine much larger series of specimens than have hitherto been available, thanks to the kindness of a number of correspondents who have sent me Char, Powan, Pollan, Cyprinoid hybrids, etc., and during the last few years I have described several new forms of considerable interest.

But much still remains to be done; I am still uncertain as to the occurrence of Pollan in Corrib and Killarney; I know that Char inhabit a number of lakes from which I have not been able to get them, and I am sure that a thorough study of the Cyprinoids, in a favourable locality like Lincolnshire or the Broads, would reveal the existence of many hybrids as yet undescribed. Moreover, growth, migrations, food, and habits offer a fine field for the student.

However, it cannot be denied that in recent years our knowledge of the fishes which live in our lakes and rivers, and of their life-history, has increased enormously, and the present time is, therefore, an appropriate one for summarizing and bringing together what is already known.

All the illustrations in this book have been drawn by myself, and it has been my aim to show the characteristic features of each species rather than to produce an artistic effect.

C. T. R.

CONTENTS

LIST OF ILLUSTRATIONS
IN THE TEXT

FULL-PAGE PLATES

INTRODUCTION

CLASSIFICATION AND TAXONOMY

Fishes defined. Classification of British freshwater fishes :
the Teleosteans. Topography of the fish · the head—the mouth
—the teeth—the gills—the pharyngeals—the scales—the fins—
coloration—the air-bladder. Measurement and counting

THE word fish is often applied in a popular sense
to any animal which lives in the water, but
zoologists use the name only for aquatic animals which
belong to the great group of Vertebrates, which have
a skull and a backbone, a brain and a spinal cord
All Vertebrates which live in the water are not
fishes, but only those which propel and balance
themselves by means of fins and obtain oxygen
from the air dissolved in the water by means of
internal gills.

The classification of the British freshwater fishes
adopted in this work is as follows :—

Class MARSIPOBRANCHII
    Order HYPEROARTII.
      Family Petromyzonidæ (*Lampreys*).
Class PISCES.
    Subclass *CHONDROSTEI*.
    Order SELACHOSTOMI
      Family Acipenseridæ (*Sturgeons*).

Subclass *TELEOSTEI.*

　　Order 1. ISOSPONDYLI.

　　　　Family Salmonidæ (*Salmon, Trout, etc*).

　　　　　　" 　　Argentinidæ (*Smelt*).

　　　　　　" 　　Clupeidæ (*Shad*).

　　Order 2. HAPLOMI.

　　　　Family Esocidæ (*Pike*).

　　Order 3. APODES.

　　　　Family Anguillidæ (*Eel*).

　　Order 4. OSTARIOPHYSI.

　　　　Family Cyprinidæ (*Carp, Roach, etc*).

　　　　　　" 　　Cobitidæ (*Loaches*).

　　Order 5. ANACANTHINI.

　　　　Family Gadidæ (*Burbot*).

　　Order 6. PERCOMORPHI.

　　　　Family Serranidæ (*Bass*).

　　　　　　" 　　Percidæ (*Perch, Ruffe*).

　　　　　　" 　　Mugilidæ (*Grey Mullets*).

　　　　　　" 　　Cottidæ (*Miller's Thumb*).

　　　　　　" 　　Gastrosteidæ (*Sticklebacks*).

　　Order 7. HETEROSOMATA.

　　　　Family Pleuronectidæ (*Flounder*).

All the fishes of the fresh waters of the British
Isles, except the Lampreys and the Sturgeons, are
Teleosteans, and have the skeleton well ossified. It
would be out of place to treat of the comparative
anatomy of the various orders and families of this
group, but it is necessary to give some account of the
topography of a Teleostean fish and some explanation
of the terms used in classification and in the distinction
of species.

In form the fish is usually moderately elongate and

compressed, deepest at or in advance of the middle of the length and tapering posteriorly; sometimes it is deep and strongly compressed, as in the stately Bream which lives in still waters, sometimes long and cylindrical, as in the Eel, which wriggles through the ooze and in and out of holes and crevices. Three regions may be distinguished—head, trunk, and tail, the gill-opening marking the limit between head and body, the vent that between trunk and tail.

The head may be naked or scaly, and the bones may be exposed or may be concealed beneath a thick skin; the eyes are typically lateral, and in front of them appear the nostrils, two on each side; the snout is the region in front of the vertical through the anterior edge of the eye, and the postorbital part of the head that behind the vertical through the posterior edge of the eye, whilst the interorbital region is that part of the top of the head which lies between the bony orbits.

The eye may be surrounded by a complete series of circumorbital bones, but the supra-orbital is absent except in a few generalized types, whilst the præ-orbital is usually the largest, and the rest, the sub-orbitals, vary greatly in their development; the region behind and below the eye is termed the cheek, and is bounded posteriorly by the præ-operculum, a bone which is sometimes crescentic, but more often angular, with a vertical and a horizontal limb; the posterior edge of the præ-operculum may be entire, serrated, or spinate. Behind the præ-operculum appear the movable opercular bones or gill-covers, comprising the operculum above, the suboperculum below and behind, and the interoperculum below and in front; below these the gill-membrane is supported by a series of branchiostegal rays, very variable in number.

The mouth may be terminal or inferior, small or large, toothed or toothless, according to the manner of life. Above the mouth two pairs of bones are articulated with the anterior part of the skull, the præmaxillaries in front and the maxillaries behind; the præmaxillaries usually meet in the middle line, but in the Pike they are widely separated, and in the Eels they are absent; the maxillary is often provided

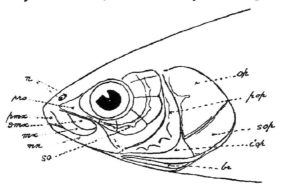

Head of Schelly (*Coregonus stigmaticus*).

Op. operculum; *sop.* suboperculum; *iop.* interoperculum; *pop.* præ-operculum; *br.* branchiostegal rays; *so.* suborbitals; *pro.* præ-orbital; *n.* nostrils; *pmx.* præmaxillary; *mx.* maxillary; *smx.* supramaxillary; *mn.* lower jaw.

with one, and in the Shad with two supramaxillary bones attached to its upper edge. In the more generalized types (Salmon, Shad, Pike, etc.) the præmaxillaries are non-protractile and are much shorter than the maxillaries, which enter the gape and may be toothed, but in many fishes the præmaxillaries nearly or quite exclude the toothless maxillaries from the border of the mouth, and the latter merely act as a lever for the protrusion of the former. The mouth is most protractile in Cyprinoids such as the Bream, and forms a sort of tube when it is protruded.

Near the mouth there may be barbels, fleshy filaments which are used as feelers when the fish is in search of food. In the Cyprinoids one or two pairs may be attached to the upper lip, and in the Loaches three or more pairs are present; the Burbot has a single barbel at the end of the lower jaw.

The mouth is toothless in the Cyprinoids, which are mainly herbivorous, and the teeth are minute in the Whitefish and Shad, which feed chiefly on minute animals, as also in the mud-eating Grey Mullets. In most cases the jaws are provided with teeth, usually conical or subconical in form; the predaceous Pike has strong erect canines in the lower jaw. The presence or absence of teeth on the bones of the palate, and their arrangement when present, are characters of considerable importance. The vomer is a bone which lies in the middle of the roof of the mouth, commencing immediately behind the junction of the præmaxillaries; it usually presents a broad anterior part, or head, and a tapering posterior part, or shaft. The palato-pterygoid arch includes five bones on each side—palatine, pterygoid, quadrate, meso-pterygoid, and meta-pterygoid; the palatines lie on each side of the vomer, and the pterygoids connect them with the quadrates, to which the rami of the lower jaw are articulated; the meso-pterygoids lie above and internal to the pterygoids, and behind them are the meta-pterygoids.

The gill-openings are usually wide, but may be restricted by the union of the gill-membranes with the isthmus, as in the Cyprinoids and Sticklebacks,

and in the Eel they are quite small.   When the gill-
covers are lifted up the four gill-arches are seen,
separated by clefts which lead into the pharynx, and
each bearing two rows of red gill-filaments ; the inner
edges of the gill-arches are usually provided with
gill-rakers, stiff appendages which vary in length and
number according to the nature of the usual food,
their object being to strain the water which has
access to the gills and to keep the food from passing
through with it.   Thus in the Pike the gill-rakers are
mere knobs, but in the Allis Shad they are close-set
bristles, very long, slender, and numerous.   The gill-
arches are supported by a special branchial skeleton,
and their gill-bearing portions are divided into upper
and 'lower limbs which form an acute backwardly
directed angle at their junction, and behind the lower
limbs of the last arch and parallel with them lie the
lower pharyngeals, a pair of dentigerous bones which
are all that remains of a former branchial arch ; these
usually bite against the toothed upper pharyngeals,
or upper elements of the preceding arches, but in the
Carp family they work against a pad on the base of
the skull, and in this group the form, number, and
arrangement of the pharyngeal teeth are of great
systematic importance.

The scales covering the body are usually arranged
in more or less regular, parallel, oblique, and longitudinal series ; as a rule they are imbricated, so that
only the posterior part of each is exposed.   In most
species they are cycloid, *i.e.* smooth and with the
edges entire, but in the Percoids they are ctenoid, *i.e.*
the edges are ciliated, as in the Bass, and in addition
the surface may be roughened with little spines, as in

the Perch and Ruffe. In the Bull-head scales are absent, in the Eel they are quite abnormal, and in the Stickleback replaced by bony scutes. The muciferous channels on the head are continued on the body as the lateral line, which appears as a series of tube-bearing scales, usually running to the base of the caudal fin; sometimes the lateral line is incomplete (Smelt, Minnow), sometimes it is absent (Shad). Each tube of a lateral line scale opens to the exterior behind, and in front communicates with a continuous longitudinal canal; this contains a series of sense organs, which are probably concerned with the perception of movements in the water.

The fins are formed of rays connected by membrane, and may be divided into median fins, comprising the dorsal, on the back, the caudal, at the posterior end of the fish, and the anal, on the lower edge of the tail between the vent and the caudal fin, and paired fins, pectorals and pelvics, corresponding respectively to the anterior and posterior limbs of other vertebrates.

The fish swims by lateral flexions of the tail, aided by movements of the caudal fin, and, as a rule, the other fins chiefly function in keeping the fish upright, or assist in turning movements, or in slowing down; however, some fishes normally progress by paddling with the pectoral fins, and in the Flat-fishes and Eels undulating movements of the dorsal and anal are of importance.

In the more generalized Teleosteans (*Isospondyli*, *Haplomi*, *Ostariophysi*) the fins are composed of flexible articulated rays, the anterior of which may be simple, whilst the majority are branched; the dorsal and anal are distinct from the caudal, the pectorals are placed low, just behind the head, and the pelvics far back, near the vent. The Eels (*Apodes*) differ in that the long dorsal and anal are confluent with the reduced caudal and the pelvic fins are absent. In the *Anacanthini* the fin-rays are articulated, but the pectoral fins are higher up on the sides, and the pelvic fins are placed below or in advance of them. In some, at least, of the *Percomorphi* and *Heterosomata* the anterior rays of the dorsal and anal and the outer rays of the pelvic fins are non-articulated spines, typically stiff and pungent, but in the dorsal fin of the Bull-head feeble and flexible; the spinous portion of the dorsal may separate off as a distinct fin. In the two last-named orders the pelvic fins are more or less advanced, and the pelvic bones are often directly attached to the clavicles. The evolution of spinous fin-rays has added a new function to the fins, that of weapons of attack and defence, and pugnacious fishes, like the Sticklebacks, know well how to use them to the best advantage.

In his interesting book on *Concealing Coloration in the Animal Kingdom*, Mr. G. H. Thayer has written : " If an object be colored so that its tones constitute a gradation of shading and of coloring *counter* to the gradation of shading and coloring which light thrown upon it would produce, and having the same rate of gradation, such object will appear perfectly flat ;— retaining its length and breadth, but losing all appearance of thickness ; and when seen against a background of color and pattern like its own will be essentially indistinguishable at a short distance."

These words give us the clue to the coloration of our British freshwater fishes, which live in a medium through which light descends from above ; consequently nearly all of them exhibit a gradation of shades from silvery white below to a dark bluish, greenish, or brownish above. Some, as for example most of the Cyprinoids, depend upon this simple shading to render them inconspicuous, but others are variously marked in harmony with the ground on which they lie or the weeds among which they lurk

A remarkable characteristic of fishes as a whole is their power to vary their colour and markings; some of the fishes of tropical seas can instantaneously change their colour, sometimes from white to black, or from yellow to red, green, or brown; just as suddenly they can become spotted, striped, or barred. This inconstancy makes coloration a very doubtful aid to the systematic ichthyologist, and among our freshwater fishes we may cite the Trout as one of the most variable in this respect, whilst the Flounder is unrivalled in its capacity for imitating the ground on which it lies, and the Bull-head is remarkable for its rapid colour changes when excited by greed, fear, or anger; such instantaneous changes are accomplished by the expansion and contraction of the variously coloured pigment cells in the skin

The air-bladder may be mentioned here because of its importance in classification; it lies between the alimentary canal and the vertebral column, and is connected with the former by a duct in the more generalized types which belong to the first four Teleostean orders enumerated above, but not in the remainder. The air-bladder may be absent (Flounder, Bull-head), or it may become connected with the internal ear by means of cæcal diverticula (Shad), or by means of a chain of ossicles which are modified elements of the anterior vertebræ (*Ostariophysi*);

in the Loaches it is reduced and wholly or partly enclosed in a bony capsule.

The usual function of the air-bladder appears to be a hydrostatic one, enabling the fish to accommodate itself to changes of pressure at different depths by the absorption or secretion of gas; what purpose is served by the connection with the auditory organ is not definitely known, but probably an increased sensitiveness to changes of pressure results.

It would be out of place in this work to give an account of the nervous, vascular, digestive, excretory, and reproductive systems, or of the internal skeleton, as it will seldom be necessary to mention them.

In the distinction of species the greatest depth of the body is usually compared with the length of the fish, measured from the extremity of the snout to the base of the caudal fin ; the length of the head is the distance from the tip of the snout to the end of the operculum. The diameter of the eye and the width of the interorbital region are usually compared with the length of the head, and the position of the posterior end of the maxillary with reference to the eye is noted. The least depth of the caudal peduncle, *i.e.* the tail just in front of the caudal fin, is compared with its length, measured from the vertical through the end of the base of the anal fin to the actual base of the middle caudal rays. The number of scales in the lateral line, or if these be enlarged the number in a longitudinal series a little above the lateral line, from above the gill-opening to the caudal fin is enumerated, and the number in a transverse series is usually reckoned from the dorsal fin to the lateral line, and thence to the base of the pelvic fins The shape, size, and position of the fins are important, and the number of rays especially so.

Males often have a larger and broader head and larger fins than females, and there may be other sexual differences, especially at the breeding season. Young fish usually differ from the adults in the longer head, larger eye, shorter snout, and smaller mouth, and the shape and relative size of the fins may change considerably during growth.

THE FRESHWATER FISHES OF THE BRITISH ISLES

CHAPTER I

THE LAMPREYS

Lampreys and Hag-fishes. The Lamprey family: the Sea Lamprey — specific characters — size — distribution — breeding —death follows breeding—structure and habits of larvæ— metamorphosis — habits of adult—origin of name—value as food. The Lampern : differences from Sea Lamprey—distribution — migrations — breeding — food and habits — uses—other names. Planer's Lamprey : differences from Lampern — distribution—metamorphosis

THE Lampreys (*Petromyzonidæ*) and Hag-fishes (*Myxinidæ*) differ widely in structure from the Selachians and true fishes, and in many respects are more generalized; they constitute a distinct class, the Cyclostomata or Marsipobranchii. They are naked eel-shaped animals without paired fins, and with the median fins supported by internal carti-laginous radials, but lacking dermal fin-rays; the internal skeleton is cartilaginous, and there are no bones anywhere; the terminal mouth has no jaws, but there is a very muscular protrusible tongue, which is armed with cuspidate horny laminæ, and works

like a piston, rasping off the flesh of the fishes on
which these animals prey; further characteristic
peculiarities of the group are the presence of only
a single median nasal opening and the structure of
the gills, which are parallel vascular laminar ridges
on the inner walls of a paired series of well-separated
muscular pouches, communicating internally with the
pharynx and opening either directly or by means of
ducts to the exterior.

Water is often both taken into and expelled from
the branchial sacs through their external apertures;
this is accomplished by the alternate expansion and
contraction of their muscular walls, and the branchial
skeleton, which is well developed only when the
sacs are large, lies external to them, and is nearly
rigid, serving to prevent the collapse of the
branchial region; this method of respiration is no
doubt due to the parasitic habits of the Marsipobranchs,
for when they are attached to other fishes they
cannot take water through the mouth into the
pharynx, and by contracting the latter expel it through
the gill-openings, which is the normal method of
respiration in other fishes.

The Hag-fishes are exclusively marine, and many
of them occur in deep water; they are in most
respects more degenerate than the Lampreys ; they
bore right into the fishes which they attack, and
devour them until practically nothing is left but
skin and bone.

The Lampreys are found on the coasts and
in the rivers of all temperate regions; there are
about twenty species, most of which spend part
of their life in the sea, whilst all appear to spawn
in fresh water. In them the mouth is surrounded

by an expanded circular lip or 'suctorial disc' which bears horny teeth; at its inner edge, and placed above and below, are two cuspidate horny plates, which may be regarded as enlarged or coalesced teeth, and are termed the supra-oral and infra-oral laminæ; within the mouth are seen the laminæ that arm the tongue, an anterior unpaired one and a posterior pair. The eyes are subcutaneous, but fairly conspicuous, and on top of the head, between or a little in advance of the eyes, appears the nasal opening; the small gill-openings, seven in number on each side, form a series which begins not far behind the eye. The fins are two dorsals and a caudal; the former are placed behind the middle of the length of the fish, and the latter is composed of a long lobe both above and below at the end of the tail.

With their suctorial disc the Lampreys attach themselves to fishes, which they devour, sucking the blood and scraping off the flesh by means of the toothed laminæ on the tongue. Whilst there can be no doubt that in many respects they are a very primitive group, many of their peculiarities are due to their manner of life, and they may even have evolved from ancestors which had jaws. All have a remarkable life-history, as the larvæ differ greatly from the adults in structure and habits, and undergo a fairly rapid transformation into the adult form after some years of larval life.

The British species are three in number, namely, the Sea Lamprey (*Petromyzon marinus*), the Lampern or River Lamprey (*Lampetra fluviatilis*), and Planer's or the Brook Lamprey (*Lampetra planeri*). The larvæ, known as Pride, Mud Lamprey, or Blind

Lamprey, have been described as a distinct genus and species under the name *Ammocœtes branchialis*.

THE SEA LAMPREY (*Petromyzon marinus*) is distinguished from the others by its dentition. The sucking disc is covered with conical teeth arranged in oblique radiating series, four inner teeth on each

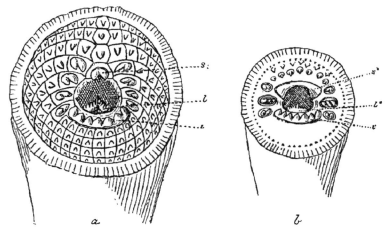

FIG. 1.—Suctorial discs of Sea Lamprey (*a*) and Lampern (*b*).
s. supra-oral lamina; *l.* anterior lingual lamina; *i.* infra-oral lamina.

side being enlarged and bicuspid; the supra-oral lamina is narrow and bicuspid, the infra-oral lamina has seven to nine cusps, and the anterior lingual lamina bears two curved denticulated ridges separated by a median longitudinal groove. The first dorsal fin is separated from the second by an interspace, the second from the caudal only by a notch. The colour is grey, yellow, or green, marbled or spotted with olive, brown, or black.

The Sea Lamprey attains a length of more than

3 feet and a weight of over 5 lbs., its usual size being from 20 to 30 inches; one of 20 inches is figured (Pl. I, Fig. 1). It is found in the North Atlantic and Mediterranean, entering rivers in Europe and North America. It is generally distributed in the British Isles, and appears to enter our southern rivers at an earlier date than those in the north, for in the Severn at Worcester the fishing season used to be from February to May, and the spawning was over in June, whereas in Scotland the Lampreys are said not to ascend until June, spawning in July or August.

By far the best accounts of the life-history of this species have been given by two American authors, Professors S. H. Gage and H. A. Surface, who have studied it in the lakes of the State of New York. Although structurally similar to the true Sea Lamprey, the Lampreys of this region are peculiar in that they never go down to the sea; they spawn in June, leaving the lakes and seeking out the clear brooks, often stealing a ride by fastening on to large fishes bound in the same direction; at this time the ground-colour changes from grey to bright yellow, and the two sexes become markedly different in appearance, the male developing a ridge along the middle of the back and a prominent anal papilla, and the female a so-called anal fin.

The Lampreys migrate independently, and a large majority of the earliest to enter the streams are males; the spawning-place is chosen where the stream is fairly rapid, usually just above a ripple, and where the bottom is sandy but strewn with pebbles. Here a space is cleared by moving the stones a little way down stream until a sandy 'nest' is formed, usually oval in form and slightly hollowed out, with a pile of stones just below it. If the male arrives unaccompanied he starts operations by himself, but on joining him the female helps; they fasten on to the stones with their suckers, and with powerful efforts loosen them and drag them down stream.

The female now secures herself by means of her sucker to some large stone near the upper end of the nest, and her mate attaches himself to her in the same way near her head, and winds himself partly round her; then the two together stir up the sand with vigorous movements, whilst the eggs and milt are simultaneously deposited. The eggs are covered with an adhesive substance, and particles of sand stick to them, so that they sink to the bottom of the nest. The pair now separate and at once commence removing stones from above the nest and enlarging the pile at the lower end, the sand thus loosened being carried down and covering all the eggs. This process is repeated at short intervals until the spawning is completed, and then they leave, many Lampreys being found going down stream or attached to stones below the nesting-places. According to Professor Surface, all die after spawning; the intestine has atrophied, they are emaciated, their skin is torn off in many places, they become stone-blind and covered with fungus, and are so completely debilitated that recovery is out of the question.

The development of the eggs is rapid, and in from ten to fifteen days the larvæ are hatched; about a month later, at the end of July, when they are about half an inch long, they leave the nests, where they

PLATE I

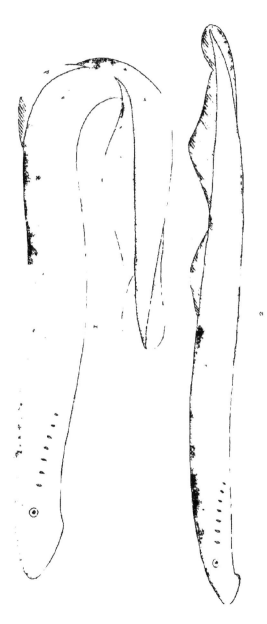

1 Sea-Lamprey : 2 Lampern

have hitherto lived in the sand, and wander down stream until they find some suitable spot, often at a bend where the water runs slowly, where they burrow in the sand or mud.

The larvæ, or Prides, as they are called in England, differ greatly from the adults; they are toothless, with a small transverse lower lip and a hood-like upper lip, and the entrance to the mouth is guarded by a number of fringed barbels which form a very perfect sieve. The eyes are hidden and quite rudimentary, the small gill-openings lie in a groove, and the fins are low.

Each Pride forms a tube in the sand or mud in which it lies; the tube is usually open at the end, so that the water flows in freely. When disturbed the Prides worm their way through the sand and do not wriggle out if they can help it; they feed on minute organisms. They probably live in this way for three or four years, and when they are from 4 to 6 inches long change into the adult form.

It may here be mentioned that the Prides of the different species of Lampreys are so similar that the name *Ammocœtes branchialis* is applicable to all. The name "Blind Lamprey" refers to their rudimentary eyes, and that of "Mud Lamprey" to their habit of remaining buried in the mud, whence they may be dug out in quantities in small brooks if the stream above be dammed up.

The metamorphosis takes place in the autumn; first the eyes appear, and as they become more evident the mouth is contracted, the upper and lower lips join to form a single circular lip, and the fringed barbels become reduced to simple papillæ surrounding the mouth; then the lips grow out to form the sucking-disc, the teeth develop, the branchial groove disappears, etc. Numerous important internal changes have accompanied those more easily observed, and the whole process occupies the greater part of two months, from about the end of August till the middle of October. Until the transformation is complete the Lampreys may still be captured in the sand, but they are more active than the larvæ which are not changing, and attempt to swim away when disturbed. After the metamorphosis the young Lampreys go down to the lakes, where they live on the fishes; sometimes a fish attacked has been observed to rise to the surface and turn over on one side, so that the head and gills of his enemy are out of the water and he is forced to drop off.

Two or three years elapse before the Lamprey attains sexual maturity and seeks the streams to end his career in perpetuating his race.

This account agrees with what we know of the Sea Lamprey in this country, except that the sea takes the place of the lakes as the home of the adult form, and, as might be expected, a larger size is reached. In the sea this species lives chiefly on Cod, Mackerel, etc., and it is on record that large numbers attached themselves to a Basking Shark, not leaving it until it was dead. Quite recently one was landed at Buckie, which had fixed itself to the rudder of a fishing-boat and could not be detached until the boat had been beached.

In their migrations the Lampreys are said to be often carried into rivers by Salmon, and have been described as ascending rapid streams by plunging quickly forward and hastily attaching themselves to some fixed object, such as a large stone, and then resting, waiting an opportunity for a new plunge; in this connection it may be mentioned that the name Lamprey is derived from the mediaeval Latin *Lampreda*, a corruption of the older *Lampetra*, from *lambere*, to lick, and *petra*, a stone.

At one time Lampreys were much appreciated as a delicacy, so much so that there was a regular fishery in the Severn, which was, and still is, the most noted river for them, although they now appear to be scarcer than formerly. The city of Gloucester used to present the reigning monarch with a dish of Lampreys every Christmas, and the death of Henry I. was attributed to a too hearty meal of these fishes.

THE LAMPERN or RIVER LAMPREY (*Lampetra fluviatilis*) differs from the Sea Lamprey especially in the dentition (Fig. 1). The disc is smaller, and has an outer series of small marginal teeth, within which is in front an irregular double row of conical teeth, and on each side three large teeth, the first and last of which are usually bicuspid, the middle one usually tricuspid. The supra-oral lamina is broad and has a pointed projection at each end, the infra-oral lamina has from six to nine more or less acutely pointed cusps, and the anterior lingual lamina bears a single transverse denticulated ridge with an enlarged median denticle. The first dorsal fin is separated by an interspace from the second, which is triangular in form and continuous with the caudal. The coloration is silvery white, with the back bluish or greenish.

This is a smaller species than the preceding; the example figured (Pl. I. Fig. 2) is a foot long, and the greatest length attained is not more than 16 inches; as compensation for this, the Lampern is much more abundant than the Sea Lamprey. It is found on the coasts and in the rivers of Europe, Siberia, Kamchatka, and Western North America from Alaska to California, and is plentiful enough in many British and Irish rivers, particularly the Severn, Trent, Ouse, Dee, etc.

Our knowledge of the life-history of this species is rather unsatisfactory, but it seems that a number of individuals reside permanently in fresh water, whilst the rest, like the Sea Lamprey, spend the greater part of their adult life in the sea and enter the rivers chiefly in order to spawn. The latter enter the rivers in the autumn or winter months, and later on make their way into the smaller streams, from April to June assembling in groups of thirty or forty on shallow fords where the bottom is gravelly; their breeding habits are precisely similar to those of the Sea Lamprey, except that several couples share in making and using a nest The larvæ cannot be distinguished from those of the Sea Lamprey, and undergo a similar metamorphosis; a considerable proportion of the young Lamperns make their way to the sea, but in some of our rivers, such as the Trent, and in lakes like Loch Lomond, a number of them remain and spend their whole life in fresh water.

Thus whilst the Sea Lamprey parallels the Salmon in that both are migratory species, living in the sea and ascending the rivers to spawn, and only losing their migratory habits where large areas of fresh water offer abundance of food, as in some lakes of Eastern North America, the Lampern and the Trout resemble each other in that they are the smaller species, including both migratory and permanently fluviatile individuals in our waters. The Lamperns attach themselves to other fishes, sucking the blood and devouring them; they are also said to feed on worms, insects, etc. In Loch Lomond they have been described as attacking and destroying the Trout and Powan; the latter especially fall easy victims to their voracity, and may frequently be seen floating on the surface of the water with the Lamperns attached, their jaws and heads buried in the flesh of their victims. In the Baltic they follow the shoals of Herring in the summer-time, but enter the rivers in the autumn.

Lamperns may often be seen swimming like Eels against the stream, or holding on to stones at the bottom; the Irish naturalist, Thompson, once observed, on a warm summer's day, a number of Lamperns attached by their mouths to the under surface of the leaves of water-lilies in a pond near Belfast, the wriggling of their dangling bodies producing a strange effect.

The Lampern can live out of the water for days, and is thus easily transported alive in baskets; it is said to be excellent food either stewed or potted, but its chief use in this country appears to be as a bait for Eels, Cod, Turbot, etc., its especial virtue being its toughness on the hook. For this purpose it used to be captured in large numbers in the Thames, and still is in the Trent, whence they are sent alive to Grimsby.

Other names for the Lampern are " Stone-eel "
and " Nine-eyes," the nasal aperture, eye, and
seven gill-openings making up the count.

THE BROOK LAMPREY or PLANER'S LAMPREY
(*Lampetra planeri*) differs from the Lampern in
that the teeth are blunter and the edge of the
suctorial disc is more strongly fringed, whilst
the dorsal fins are contiguous or continuous,
separated by a notch only, and the free edge of

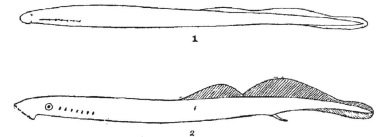

FIG. 2.—1. Larval Lamprey or Pride (*Ammocœtes*). 2. Planer's Lamprey.

the second dorsal is usually more distinctly rounded,
instead of angular.

It is a smaller species, not growing to more than
half the length of the Lampern, and has a different
distribution, for whilst both species are found in
Europe and Siberia, this one occurs also in Japan,
but is absent from America. It is common in
Britain to at least as far north as Perthshire, and
is widely distributed in Ireland.

Planer's Lamprey is usually found in the smaller
streams, brooks, and ditches, and never goes down
to the sea. It was in this species that the transformation of the larval Pride into the adult

Lamprey was first observed. Professor A. Muller, in 1856, described how he had watched Planer's Lampreys spawning in a brook near Berlin and had seen the eggs hatch out in about three weeks and develop into Prides, previously known as *Ammocœtes branchialis.* According to his further observations, the larvæ live three or four years before they change into the adult form, the transformation occupying twenty-six days or more, and then they live only for a few months, dying after spawning. The breeding habits of this species are precisely similar to those of the Lampern, and the season for spawning is the same.

CHAPTER II

THE STURGEON

The class *Pisces* the subclass *Chondrostei*—palæozoic Chondrosteans—the Paddle-fishes The Sturgeon family the Sturgeon described—variation in number of scutes—only one British species—distribution—size—food and habits—only stragglers in our rivers—commercial importance—names

WITH the exception of the Lampreys all the freshwater fishes of the British Isles belong to the class *Pisces*. In this class the body is usually covered with scales, which in the more generalized types have the form of juxtaposed rhombic bony plates, but in the more specialized ones are thin, rounded, and imbricated. Median and paired fins are present; they are supported basally by an internal skeleton, and are also provided on each side with a series of dermal rays. The endoskeleton may be cartilaginous or bony, but there are always a number of external or membrane bones; the mouth is bordered by bony jaws; the pectoral arch is usually connected with the skull by a series of bones The gill-sacs of the Lampreys are represented by clefts leading from the pharynx into the branchial chamber, which opens behind to the exterior, and is covered by the opercular membrane, supported by the opercular bones and branchiostegal rays; the interbranchial septa, or walls of the clefts, are so reduced that the

14

gill laminæ project outwards as filaments, and the branchial skeleton, supporting the pharynx between the clefts, is movable; water is taken in through the mouth, and by the contraction of the pharynx is sent through the gill-clefts and over the gills, and so outwards *via* the gill-openings. The nasal sacs are paired and lateral; each is usually provided with two external nostrils. An air-bladder is typically present.

Of four subclasses of the *Pisces*, two, the Crossopterygians and Dipnoans, are represented at the present day by only a few species in the fresh waters of tropical countries. The remaining two, the Chondrosteans and Teleosteans, include British freshwater species.

The Sturgeons (*Acipenseridæ*) and the Paddlefishes (*Polyodontidæ*) are the specialized and in many ways the degenerate living members of a very ancient and generalized group of fishes The essential characters of the subclass *Chondrostei* to which they belong are as follows:—

The bones overlying the primary pectoral arch and connecting it with the skull are four in number; the rays of the dorsal and anal fins are much more numerous than their endoskeletal supports, which they simply overlap; the pelvic fins, like the dorsal and anal, have each a muscular basal lobe containing a well-developed series of radial segments of their skeletal rods; the caudal fin is heterocercal, *i e.* the axis of the tail is turned upwards, the upper (supra-caudal) lobe is reduced to a series of spines, and the lower (infra-caudal) lobe is strongly developed.

In the Chondrostean fishes found fossil in the palæozoic rocks the mouth, head, and scaling were normal, but their modern descendants are very remarkable. The Paddle-fishes (*Polyodontidæ*) of China and the Mississippi have the body naked, the eyes very small, the snout much produced, and the

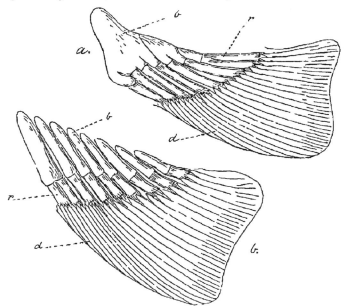

FIG. 3.—Pelvic fin (*a*) and anal fin (*b*) of a Sturgeon.
b. basal cartilages ; *r.* radials ; *d.* dermal rays.

mouth rather large and non-protractile; there are no præmaxillaries and no branchiostegal rays. The Sturgeons (*Acipenseridæ*), with twenty or more species from the seas and rivers of Europe, Asia, and North America, differ from them especially in the presence of a transverse series of four barbels on the lower side of the snout, in front of the small, protractile mouth, whilst there are five longitudinal rows of large bony scutes on the body, a mid-dorsal and paired lateral and ventral series; the primitive scaling is retained on the upturned part of the tail; another peculiarity is the presence on each side of a strong pectoral spine, formed by the fusion of the anterior fin-rays.

The Sturgeons are mostly anadromous fishes, living in the sea and entering rivers to spawn, but in some of the large lakes and rivers of North America they have become permanent residents. They are very variable, change greatly during growth, and also hybridize to a considerable extent, so that there has been considerable difference of opinion as to the number of species entitled to recognition; it is generally agreed that there is only one species in Western Europe, but some authors are of opinion that a second species may occasionally cross the Atlantic from America and enter European rivers, a matter which is dealt with below.

The Sturgeon (*Acipenser sturio*) has the body elongate, subcylindrical anteriorly and tapering posteriorly. The size, shape, and arrangement of the bones which cover the upper surface of the head are subject to great variation; thus the frontals may be in contact for the whole of their length, or may be completely separated by an ethmoidal plate. The snout is covered above with small bony plates and is rather long; in young specimens it measures at least half, and in examples of from 2 to 3 feet not much less than half of the length of the head, but in large fish only about one-third; the insertion of the barbels is nearly equidistant from the mouth and the tip of the snout in young and half-grown

2

examples, but is nearer to the end of the snout in large fish. The dorsal fin is placed far back, and is composed of from thirty-seven to forty-six rays; the anal, which begins below the middle of the dorsal and extends a little farther back than that fin, has twenty-four to twenty-nine rays, and the pelvic fins have the same number. There are from ten to sixteen scutes in the dorsal series between the head and the dorsal fin, and from nine to fourteen in each ventral series between the pectoral and pelvic fins; the lateral series increases in number to some extent with age, owing to the addition of small scutes posteriorly; in 1870 Dr. Günther counted the lateral scutes on a number of European examples, and found twenty-six to thirty-one in the young, but twenty-nine to thirty-four in half-grown and adult fish. A larger series of specimens would no doubt have shown greater variation; thus Day found only twenty-seven scutes on one side and twenty-eight on the other in an adult fish more than 5 feet long, from Margate, whilst in a specimen 18 inches long, from the Adriatic, I count thirty-four on one side and thirty-five on the other and in one of 30 inches, from the Gironde, I count forty on each side, and after careful comparison I am convinced that these are true *Acipenser sturio*. It has been assumed that specimens with a large number of lateral scutes, as, for example, one said to be from the Tay, with thirty-eight on one side and forty on the other, belong to an American species (*A. maculosus*), which has been supposed to cross the Atlantic occasionally; probably they cannot be distinguished from American examples named *Acipenser maculosus*, but I am unable to separate these specifically from other American specimens with fewer scutes, named

Acipenser sturio. In young examples the scutes are juxtaposed and each is furnished with a strong compressed recurved spine, but with age the scutes become separated and the spines disappear. The coloration is usually of a greyish or brownish purple on the back, replaced by white below the lateral series of scutes; small dark spots may be present in the young.

The Sturgeon is an inhabitant of the coasts and rivers of Eastern North America and of Europe from Scandinavia to the Black Sea; it is most abundant southwards and prefers larger rivers than any we can offer; although it is by no means a rarity in our islands, only solitary stragglers are usually encountered.

It is a very large fish, attaining a length of 18 feet; a specimen from Heligoland, more than 11 feet long and of 623 lbs. weight, was sent to Frank Buckland, and records of British examples of 7 or 8 feet are too numerous to mention. Our figure (Pl. II, Fig. 1) is of a quite young fish, 12 inches long.

The Sturgeon is a somewhat sluggish fish, which feeds chiefly on small invertebrates, rooting up the sand or mud with its snout and feeling for its prey with its barbels; in the spring and early summer it enters the rivers to spawn, and in the countries where Sturgeons are plentiful and ascend the rivers in large companies, a profitable fishery is carried on for this and other species. In the British Isles there is no regular fishery, but Sturgeons are occasionally taken in Salmon nets; they certainly do not enter our rivers in any numbers, and probably do not spawn in them, although they may ascend for some distance, specimens having been captured in the Severn at

Shrewsbury and in the Trent at Nottingham. According to Macpherson, a Sturgeon recently captured may be a dangerous companion, and one has been known to cut a man's leg to the bone with a blow of its tail, the dorsal spines and the sharp-edged lateral scutes making the latter a formidable weapon of offence.

The flesh of the Sturgeon is generally esteemed, and the preparation of caviare from the roe, and of isinglass from the air-bladder, of this and other species are large industries, principally carried on round the Black and Caspian Seas.

The French name for this fish is *Esturgeon* and the German *Stor*, whilst similar words occur in other languages and are probably from the same root as the verb 'to stir,' in allusion to the habits of the fish.

CHAPTER III

THE SALMON

The Teleosteans. The Salmon family: its marine origin—
the British genera. The Salmon distribution — differences
from Trout—life in the sea—ascent of rivers—feeding in fresh
water—characters of breeding fish—spawning—Kelts—Rawners
—Hybrids — Alevins — Parr — Smolts — Grilse — small spring
Salmon—maiden Salmon—annual Spawners. Age of large fish
—characters of large fish—the homing instinct—names of
Salmon at different periods of life defined—Dahl's researches—
marking experiments—scale investigation. Ripe Parr—land-
locked Salmon—enemies of Salmon—lice and maggots
—disease—Salmon as food—ancient fisheries—litigation caused
by the Salmon—improvement of Salmon rivers—Bibliography
—the Salmon of the North Pacific

ALL the remaining British freshwater fishes
belong to the class *Pisces* and the subclass
Teleostei, and have a well-ossified skeleton; they differ
from the *Chondrostei* as defined on p. 15, in
several features of specialization; the bones which
overlie the primary pectoral arch and connect it with
the skull, are reduced in number to three; the
dorsal and anal rays are less numerous, and are
usually not separable into lateral series, each being
united in the middle line with the corresponding ray
of the other side; the muscular lobes at the bases of
the fins and their included series of radial segments
of the skeletal supports are greatly reduced, and

each ray is now attached to its own support; the
pelvic radials also are small or absent, and some or
all of the rays are directly inserted on the pelvic
bones; finally, the caudal fin is 'homocercal,' *i.e.*
the upturned end of the tail has become so short
that the fin is mainly formed of rays which were
originally inferior, but are now terminal.

The more generalized Teleostean fishes may be

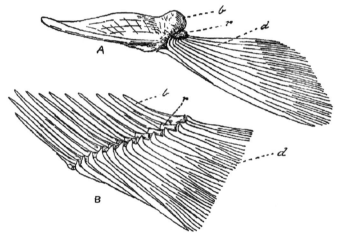

FIG. 4.—Pelvic fin (A) and anal fin (B) of a Trout.
b. basal bones; *r.* radials; *d.* dermal rays.

characterized by the terms, *Physostomi, Malacopteri,
Abdominales.*   That is to say, the air-bladder communicates with the alimentary canal by means of a
duct, the fin-rays are flexible and jointed, and the
pelvic fins, when present, are placed far back.

Of several orders possessing these features four are
represented in the rivers of the British Isles, namely,
the *Isospondyli*, including the Salmon family (*Salmonidæ*), the Smelt family (*Argentinidæ*), and the Herring

family (*Clupeidæ*); the *Haplomi*, containing the Pikes
(*Esocidæ*); the *Apodes*, including the Eel family
(*Anguillidæ*), and the *Ostariophysi*, wherein are placed
the families of the Carps (*Cyprinidæ*) and the
Loaches (*Cobitidæ*).

These will now be treated of in the sequence indicated, and will then be followed by the more

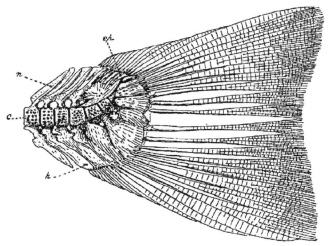

Fig. 5.—Caudal fin of a Trout, showing the upturned posterior
  vertebræ and the rays supported by the hypurals or hæmal
  spines (*h*).

  c. centrum ; *n.* neural spine ; *ep.* epurals (supra-caudal basalia).

specialized orders, terminating with those fishes
which have lost the pneumatic duct and have
developed spinous fin-rays, and in which the pelvic
fins have migrated forwards to a position below or in
advance of the pectorals.

The Salmon and its allies (*Salmonidæ*) are perhaps
the most interesting of all the groups of fishes ; they
are of great economic value, and the sportsman will

certainly place them first on account of their size, beauty, and activity, whilst the naturalist finds endless scope for work in the solution of the problems connected with their life-history and distribution, and in the attempt to distinguish and define the numerous races and species

The Salmonidæ are fishes with a naked head and scaly body, which in form and appearance are more or less similar either to a Trout or a Herring. The non-protractile mouth is bordered above by the præmaxillaries in front and the maxillaries at the sides ; the fins have no spinous rays ; the pectorals are placed low and the pelvics far back, below the dorsal, which is in about the middle of the length of the fish ; the presence of an adipose fin, a small fleshy flap on the back above the end of the anal fin, is an important character common to all the members of this family, whilst the absence of oviducts is a peculiarity worthy of mention.

The Salmonidæ are found in the arctic and temperate regions of the Northern Hemisphere, and may be regarded as marine fishes which are establishing themselves in fresh water ; there are many permanently fluviatile or lacustrine forms, but a number of others spend a considerable part of their life in the sea, and in some cases non-migratory and migratory fish occur within the limits of one species. Some writers look upon the Salmon and its relatives as true freshwater fishes which have acquired the habit of going to the sea for food, and which return to their original home to spawn. Against this it may be urged that whereas many marine fishes take to fresh water, the reverse is a rare phenomenon ; to illustrate the former we may instance the Bass

PLATE II

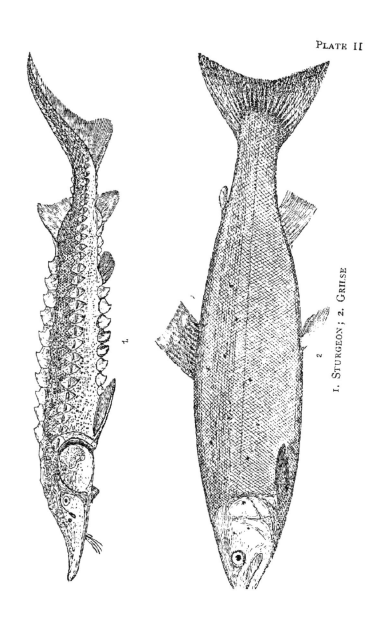

1. STURGEON ; 2. GRILSE

(*Morone*), which belong to the marine family of the Sea Perches (*Serranidæ*); two species live in the sea or in estuaries, two more ascend rivers to spawn, and two have become permanent residents in fresh water; also, the Shad are fishes of the Herring family which spawn in rivers, or may even spend their lives in fresh water.

However, this is only argument from analogy, and more positive evidence is derived from the distribution of the species, which affords abundant proofs of their marine origin; the Salmon is not a European but an Atlantic species, the Trout extends eastwards to the Caspian and Aral Seas and their tributaries, but is absent from the Siberian rivers to the north of them, wherein our Roach, Pike, and Perch are found; with one or two exceptions, probably due to recent geographical changes, the Char and Whitefish of the Alps are found only in those lakes which do not communicate with the Mediterranean. In short, the distribution of the Salmonoids has been mainly determined by the seas to which they resort, or those from which they formerly came.

The British members of this family belong to four genera, namely, *Salmo* (Salmon and Trout), *Salvelinus* (Char), *Coregonus* (Whitefish), and *Thymallus* (Grayling), the salient differences between which are shown in the following synopsis:—

I. Mouth rather large, the maxillary extending at least to below the middle of the eye; teeth well developed; scales small or moderate; dorsal fin with 10 to 16 rays.

    No depression behind the head of the vomer, and a double or zigzag series of teeth present on the shaft of that bone, at least in the young—Salmon and Trout.

Vomerine teeth restricted to a group on the head of the bone, behind which there is a boat-shaped depression—Char.

II. Mouth small, the maxillary short; scales of moderate size.

Teeth minute or absent; dorsal fin with 10 to 16 rays—White-fish.

Teeth small; dorsal fin with 18 to 24 rays—Grayling.

The Salmon (*Salmo salar*) is an inhabitant of the North Atlantic, ranging northwards to Hudson Bay, Greenland, Iceland, and Northern Europe, and southwards to Cape Cod and the Bay of Biscay, and entering all suitable rivers to spawn. In the British Isles it is most abundant in Scotland and Ireland; Wales and the West of England come next, but the pollution of the Thames and some other English rivers renders them uninhabitable for this species at the present day; in former times the Thames was a noted Salmon river, but even in 1841 Yarrell wrote, " A Thames Salmon is a prize to a fisherman, which like other prizes, occurs but seldom. The last Thames Salmon I have a note of was taken in June 1833."

The adult fish has a slender fusiform body and a rather small conical head; it is silvery, bluish grey above and white below, with scattered blackish spots present only above the lateral line, except sometimes anteriorly; the dorsal, caudal, and pectoral fins are bluish black, the first named often with some blackish spots near the base, whilst the pelvic and anal fins are white. Of characters which distinguish Salmon from Trout we may note that the dorsal fin has usually more rays (10, exceptionally 9, to 12 branched rays in the Salmon, 8 to 10, exceptionally 11, in the Trout), and that the scales on the tail are larger, in an oblique series from the posterior edge of the adipose fin downwards and forwards to the lateral line numbering 10 to 13 in the Salmon, 13 (exceptionally 12) to 16 in the Trout; these figures are based on an examination by the author of more than a hundred examples of each species, and these numerical differences are of special importance because they are not subject to change with the growth of the fish. In a Salmon the tail is more constricted at the base of the caudal fin than in a Trout, and consequently the anterior caudal rays form more of a shoulder, so that a Salmon does not slip through the fingers when grasped round the caudal peduncle, but a Trout usually does; the caudal fin is more or less emarginate, or in large specimens truncate, or even rounded, but is usually more distinctly notched than in Trout of the same size. In adult Salmon the anal fin is less pointed than in Trout, so that when it is laid back the last ray usually extends farther than the longest, the reverse obtaining in the Trout. The maxillary extends to (in Grilse or small Salmon) or a little beyond (in large fish) the vertical from the posterior margin of the eye, being shorter than in the Trout.

Salmon live in the open sea, where they feed on Herrings, Mackerel, Sand-eels, etc.; they spawn in fresh water from September to February, and throughout the year they approach the coasts and enter the rivers, in the estuaries often ascending with the tide, and afterwards displaying great strength and perseverance in making their way up stream, leaping falls and swimming up rapids.

During their ascent they often lie in large shoals in certain pools, or they may rest behind big stones which break the current. Opinions differ as to the height which a Salmon can clear at a single spring, but most authorities think 10 feet about the limit, and direct falls of more than this height are impass-able. Fish have been known to make repeated vain attempts to clear falls which are too high for them, only giving up at last through sheer exhaustion. It is from this habit that the Salmon takes its name, the Latin *Salmo* being from the same root as *salire*, to leap.

It is only in the larger rivers that there are regular runs of clean Salmon in the winter and early spring, and it seems that in some of the Scottish rivers, provided that the temperature be not too low or the obstacles to be overcome impassable at this season, a good proportion of these fish ascend fairly rapidly right up to the head waters, where they stay without actively feeding until the following autumn, when they spawn there. In Ireland it has been ascertained on the Blackwater that some of the winter clean Salmon make only a temporary stay, and in the spring drop back and resume active feeding in the sea, to reascend later in the season. This probably happens elsewhere in the case of many of the fish which run up early in the year, but as a general rule it may be stated that once a Salmon has entered fresh water he usually stays there until he has spawned. The summer fish run up more quickly than the spring ones; they are not kept in the lower reaches by cold floods, the water is lower, and the falls and rapids are less difficult to pass; besides, the spawning season is nearer.

The vexed question whether Salmon feed in fresh water may now be regarded as settled in the negative; that is to say, there is no active pursuit of food or regular feeding with resultant increase of weight; indeed, from the time a Salmon enters fresh

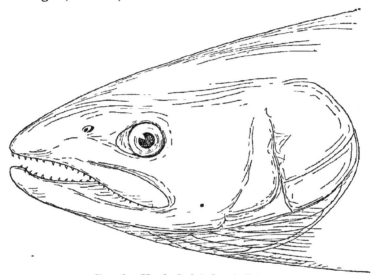

FIG. 6.—Head of adult female Salmon.

water its weight gradually decreases. It is true that occasionally Salmon may take the fly which is on the water, or that some small fish or other tempting morsel which comes within easy reach may be snapped up, but it is by no means certain that the various lures with which the angler tempts the Salmon really appeal to his appetite. The present writer, when spinning for Perch with a small spoon or a Devon Minnow on the River Yeo, has sometimes caught as many as three or four Roach in one day with this bait, and in the clear water has seen them follow it for quite a long distance before seizing it ; now, nothing is more certain than that Roach do not as a rule chase and eat Minnows, and it is probable that excitement or curiosity led these fish to their fate, and may also have done the same for many a Salmon.

When Salmon first run up, fresh from their abundant feeding in the sea, they are usually in high condition, and present the graceful form and brilliant silvery appearance with which all are familiar. The flesh of these fresh-run fish is firm and red, and there is a large store of fat in the tissues. As the spawning season approaches the fat becomes expended on the development of the sexual organs, and the flesh becomes pale and watery ; at the same time considerable changes in external appearance take place ; the silvery coloration is replaced by a dull reddish brown tint, and in the males the front teeth are enlarged, the snout and lower jaw are prolonged, and the latter is hooked upwards at the tip (Pl. III) the skin of the back becomes thick and spongy, so that the scales are imbedded in it, and large black spots edged with white appear on the body, which is also spotted and mottled with red and orange ; such Salmon are termed " red fish," and the females, which are darker than the males, are known as " black fish."

A sojourn in fresh water is not necessary for the ripening of the sexual products, for Salmon continue to run up during the spawning season, and quite ripe Salmon, with the characteristics just described, may enter from the sea and at once spawn on reaching the heads of the estuaries.

PLATE III

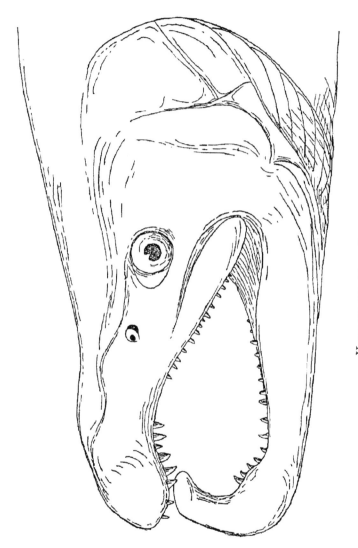

HEAD OF OLD MALE SALMON

3

Captain Barrett Hamilton has made an interesting attempt to account for the differences between the sexes at this season. His view may be shortly presented thus: The rapid transference of substance from the muscles to the genital glands is a complicated process which leads to the formation of more by-products than can be excreted, and these give rise to the growths at the ends of the jaws, or appear as pigments in the skin. The female organs are the larger, and make use of more kinds of material; to take only one instance, the red and yellow pigments of the flesh are in great part transferred to the ovaries, whereas in the males they are nearly all deposited superficially. The male organs are smaller, and their development entails more waste; hence the long hooked jaws of the males.

Salmon about to spawn select a place where the stream flows fairly rapidly over a gravel bottom; here the female scoops out a shallow trough by moving her tail from side to side, and sinking into it deposits some eggs, which are fertilized by the male fish who is waiting near; she then covers the eggs over with gravel by similar strokes of her tail, burying them at a depth of about a foot, the whole process is repeated at intervals of a few minutes, the fish gradually moving up stream, so that when all the eggs are shed, which usually takes from a week to a fortnight, the spawning-bed or 'redd' of one pair of fish may be several feet long. At this time the old male Salmon are sometimes very fierce, driving away any others who may approach them, and even engaging in desperate combats.

The spent fish are termed "kelts" or "slats," and are easily recognized by their large heads and by their leanness. In all rivers grilse kelts do not lose much time in getting back to the sea, and in small rivers this is true of all kelts; but in the larger rivers the kelts are wont to linger in the deep pools, especially big fish of the female sex, which may not reach the sea until the early summer. The hungry appearance of the kelts, and the fact that during their stay in the rivers they seem to improve in condition, the scales becoming more or less silvery, formerly led to the belief that they fed ravenously, and they were particularly credited with cannibalism, and were said to devour numbers of parr. However, it now appears that kelts do not feed until they reach the sea, or at least the estuaries, and that the improvement in a so-called " mended " kelt is mainly superficial, and consists of the assumption of a silvery livery and the reduction of the prolongations of the jaws by absorption of the connective tissue which forms them.

A good many kelts never reach the sea, especially if the journey thither be at all an arduous one; in their enfeebled condition they are very susceptible to disease, readily succumb to any injuries they may sustain, or fall an easy prey to poachers, otters, or other enemies.

According to Mr. Malloch, ripe Salmon may occasionally miss spawning, and in the spring become silvery like the kelts, when they are termed "baggots" or "rawners"; the subsequent history of these fish is obscure.

When Salmon are on the spawning-beds they are often attended by male Trout, which seize any opportunity which presents itself, such as the temporary absence of the male Salmon in pursuit of an intruder, to shed their milt on the ova; hybrids may thus be formed, but they would be extremely difficult to recognize; it may be, however, that certain examples which occasionally occur, and which it is not easy to refer with certainty to either species, are really hybrids. It has been ascertained experimentally that the hybrid off-spring of Salmon and Trout are deficient in vitality, often malformed, and seldom, in the case of the males probably never, come to maturity; however, eggs obtained from ripe female hybrids were milted from a Lochleven Trout, and a large proportion of them hatched, and the young fish did well.

The eggs hatch out at the end of the winter, and the fry, or *alevins*, weighed down by the large yolk-sac which contains their food for the first month or two of their lives, continue to dwell in the spaces between the stones of the spawning-bed, or "redd," until the yolk-sac is absorbed, and they come out and begin to shift for themselves; they now form shoals of little fish about an inch long, which live on the shallows, and usually attain a length of 3 or 4 inches in a year and 5 or 6 inches in two years, during which they feed and grow in fresh water, and are known as *parr*, their food consists of shrimps, insects, etc., and they rise to the fly readily.

The author has examined a number of parr from 3 to 4 inches long, taken at Romsey at the end of February, and consequently about a year old. In these the back is bluish or purplish, and there is a series of seven to eleven large, vertically elongate, oblong, or oval spots of the same hue, the parr-marks, along the middle of the side; behind the eye appear three blackish spots in a straight

line, the last two of which are on the operculum;
one or sometimes two of these may be inconspicuous.
On the body there are some blackish spots above
the lateral line, in some absent posteriorly, in others
present also anteriorly below the lateral line; there
is a red spot between each pair of pair-marks, and a
few other red spots are usually present.   The dorsal
fin usually has dark spots arranged in two series, the
caudal is dusky, the pelvics and anal pale, the
pectorals bluish black except towards the base.   The
maxillary reaches to below the middle of the eye,
the pectoral fin extends to or beyond the vertical
from the origin of the dorsal, the longest anal ray,
when laid back, reaches at least as far as the last,
the middle caudal rays are not more than three-
fifths as long as the longest, and the least depth of
the caudal peduncle is about one-tenth the length
of the fish (to the base of the caudal fin).

A number of Trout of the same size (Pl. VII,
Fig. 1) were taken with these parr, from which they
differ in that the spots are more numerous and more
widely distributed, whilst the regular series behind
the eye is absent, the pectoral fin is orange
coloured and the anal has a white anterior edge
bordered posteriorly by a dark streak.   The
maxillary extends to ·below the posterior edge of
the pupil, the pectoral does not reach the vertical
from the origin of the dorsal, the middle caudal rays
are two-thirds as long as the longest, and the least
depth of the caudal peduncle is about one-eighth the
length of the fish.

On Plate IV are figured two parr; the smaller is
shown of the natural size, but the larger, a fish of
8 inches long, is reduced, this, and other two-year-

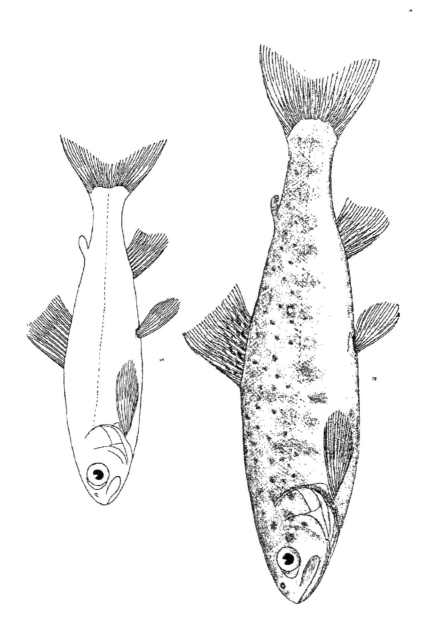

old parr, 5 to 9 inches long, in the British Museum collection, differ from those described above in that the maxillary extends to the vertical from the posterior edge of the pupil and the pectoral fin is shorter, not reaching the vertical from the origin of the dorsal.

Mr. Malloch has observed that in the winter months parr leave the shallows and rest under stones; he has often lifted a flat stone quietly and has seen three or four parr, which did not swim away for a time, but appeared dazed and sleepy.

When they are about two years old, or a little more, parr become very silvery, their parr-marks are obscured, and they are known as *smolts*, the name being derived from an old English word meaning "shining." This silvery appearance is doubtless produced by the formation of iridocytes, granules or plates of guanin, in the skin outside the scales. The smolts drop down the rivers, and on reaching the estuaries appear to migrate rapidly right out to the open sea. Although the majority of the smolts are two years old, it is possible that some may be only one, or that a few may be three years old. On some Norwegian rivers the parr are said not to change into smolts and migrate to the sea until they have spent as many as five years in fresh water.

The main descent of the smolts usually takes place about May, but varies from March to July in different seasons and on different rivers. Sometimes there is a second migration in the late summer or autumn, but Mr. Malloch thinks that this is the case only when the parr have made their way into small streams, and a deficiency of water has kept them back during the course of a dry summer.

In the open sea, as has recently been proved by a Norwegian, Dahl, the smolts pursue the young Herring, Sand-eels, etc., and this diet suits them so well that they grow rapidly, so that when they return in May or June, after spending about twelve months in the sea, to the river from which they went, they rarely measure less than 16 inches in length, or weigh under $1\frac{1}{2}$ lbs., whilst some are much larger, and those which ascend later in the season, and which have been feeding for a longer time, may weigh as much as 12 to 14 lbs.

These fish on their first return from the sea, at an age of three to three and a half years, are termed *grilse*; one of about 4 lbs. is shown on Pl. II, Fig. 2. Exceptionally small grilse are sometimes captured in February or March, but as a rule they do not appear until April, and only in large numbers in June, July, and August.

Salmon intermediate between smolts and grilse are not often captured, and mention may therefore be made of a specimen in the British Museum from Yarrell's collection, which that writer described as the smallest Salmon he had ever seen that had been to the sea. It measures only $11\frac{1}{4}$ inches in length, and was captured in the Solway Firth on the 8th of June, when it had probably been in the sea not more than two or three months.

Grilse have the coloration already described as characteristic of the adult fish, and all trace of the parr-marks has disappeared. They are of a more graceful form than the parr; the fins are relatively much smaller and the anal is differently shaped, the longest ray, when laid back, not reaching so far as the last, whilst the caudal is less strongly emarginate. In grilse the maxillary extends to, or nearly to, the vertical from the posterior edge of the eye, or sometimes slightly beyond, whilst in Sea-trout of this size the maxillary reaches back well beyond the eye.

Grilse come in from the open sea and appear on the coast in large companies, and in the summer many of them ascend the rivers from which they came, to spawn in the autumn and return to the sea as grilse kelts in the winter or early spring; on the other hand a number of them do not enter fresh water until the winter or spring, when they are nearly four years old and are known as small spring Salmon; these are very similar to grilse in size and appearance, but do not spawn until the following season. Further, it seems that quite a number of fish not only pass through the grilse stage in the sea, but may stay there until they attain a considerable weight, entering fresh water for the first time as "maiden" Salmon when they are four, five, or even six years old.

Salmon which have spawned exhibit just as great differences in the time which elapses before they return to the rivers; some spawn in successive seasons and the kelts pass only a few months in the sea before they go back again; others may miss a year, whilst others still may let two years or more go by before they again feel impelled to ascend the rivers in order to reproduce their kind.

We are now able to see that Salmon of the same weight may be of very different ages, annual spawners increase in weight very slowly, for the time spent in the sea sometimes scarcely does more than make good the loss entailed by the stay in

fresh water, and especially by the act of reproduction; exceptionally heavy Salmon are not
necessarily, or even probably, exceptionally old,
but are those which have passed the greater part of
their life in the sea, have fed well, and have spawned
seldom.

On many rivers a large proportion of the fish
spawn only once.   Mr. Malloch found on the Tay
that 80 per cent. of the Salmon which entered were
maiden fish, these are much better for the table
than those which have spawned previously; the
latter have the flesh pale and coarse, and are distinguished by the presence of numerous small spots
on the head and shoulders.   There is good reason
to believe that Salmon rarely live to be more than
eight or nine years old, or spawn more than three or
four times; on the Tay the age and history of some
large fish of from 30 to more than 40 lbs. weight
has been ascertained, and these have been either
seven or eight years old, and have spawned either
once or twice; Salmon of more than 80 lbs. weight
have occasionally been captured, and we may conclude that such fish are either maiden Salmon or
have not spawned more than once, and have grown
with remarkable rapidity.   Three maiden Salmon
from the Wye of 40, 42, and 44 lbs, mentioned by
Mr. Hutton, were six years old and had spent four
years in the sea; such fish might well have reached
a weight of 80 lbs. at the age of eight or nine years.

The Salmon of some rivers grow to a much larger
size than those of others; the Tay is a river noted
for its heavy Salmon, and the record Scottish
Salmon, a fish of 84 lbs., was taken in the Tay
estuary.   In the British Museum there is a cast of

a Tay Salmon of $55\frac{1}{2}$ lbs., and in the Buckland
collection two of $69\frac{1}{2}$ and 70 lbs., both from the
Tay, are represented by casts. Salmon also run very
large in the Tweed and in many rivers in Ireland.

These large Salmon, especially if of the male sex,

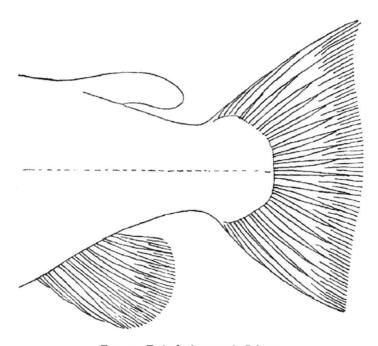

FIG. 7.—Tail of a large male Salmon

are of a much less graceful build than the grilse ;
they are often very stout, with the caudal peduncle
short and deep, but still constricted posteriorly, and
the edge of the caudal fin convex ; they are especially
distinguished from large Sea-trout by the shorter
maxillary, by the number of scales and fin-rays,

and by the form of the caudal peduncle and anal fin.

The fact that Salmon nearly always return to the rivers from which they came may perhaps be explained on the supposition that they never really lose touch with their river, that the particular feeding-grounds to which the fish of any stream resort are, as it were, within its sphere of influence. This would account for the peculiarities of form and size associated with Salmon from different rivers, for the area of the feeding-grounds would be to some extent proportional to the size of the stream Thus Salmon from a small river would seldom grow large, and, if they did, might make their way into the nearest large river instead of ascending their own.

The homing instinct is a powerful one, but may be broken through; every year a few Salmon appear off the mouth of the Thames and, if the water were pure enough, would ascend and reconvert it into a Salmon river, which will never be the case under present conditions, however many parr may be placed in the upper waters Mr. Holt instances the case of the Ballysodan River, which became a Salmon river by the successful engineering of a pass on the previously inaccessible fall at its mouth, as effectually disposing of the theory that the Salmon chooses the particular river from which it came, and no other.

The normal life-history of the Salmon has now been described, and it may perhaps be useful to offer definitions of the names applied to the fish at various stages; these are—

ALEVINS.—Fry with yolk-sac attached, found on spawning-beds in the early spring.

PARR, PINK, or GRAVELING.—Young fish with parr-marks, ranging up to 6 or 7, or, exceptionally, as much as 9 or 10 inches in length, living in fresh water until they are about two and a quarter years old, rarely more or less,

SMOLTS —Young fish preparing to descend to the sea or actually on their way there, with the parr-marks obscured by a bright silvery livery, usually about two and a quarter years old, and about 6 inches long.

GRILSE or SALMON PEAL —Fish normally from three to three and a half years old, which went to the sea as smolts in the preceding season ; rarely weighing less than $1\frac{1}{2}$ or more than 10 lbs. ; a large number of grilse enter the rivers during the summer months in order to spawn in the autumn.

SALMON —Older fish which have either spawned as grilse or have passed through the grilse stage in the sea , according to the season when they enter the rivers they are known as *Spring*, *Summer*, *Autumn*, and *Winter Salmon* ; Winter Salmon, however, may be divided into *Late Autumn Salmon*, unclean fish which spawn as soon as they enter fresh water, and *Early Spring Salmon*, clean fish which do not spawn until the following autumn.

MAIDEN SALMON —Fish which have never spawned and are entering fresh water for the first time.

DROPPERS —Winter or early spring Salmon which drop back to the sea and reascend later in the season.

FRESH-RUN FISH.—Fish which have recently entered fresh water.

CLEAN FISH.—Silvery fish with firm flesh and the sexual organs little developed.

UNCLEAN FISH.—Fish which are nearly ready to spawn, or which have recently spawned.

RED FISH.—Ripe male Salmon.

BLACK FISH.—Ripe female Salmon.

KELTS or SLATS.—Spent Salmon; if these stay in fresh water they become more or less silvery and are termed *Mended Kelts*.

Some of these names explain themselves, but the meaning of others is uncertain. Grilse corresponds to the Scandinavian *Gralax*, Grey Salmon. Parr is from the old English *parren*, to enclose, from the resemblance of the marks on the sides to the bars of a fence, and Graveling is doubtless connected with the habits of these young fish, which frequent gravelly shallows. Smolt is from the Anglo-Saxon *smeolt*, shining, in allusion to the silvery colour.

Not long ago the history of the Salmon was very obscure, but investigations made in recent years have cleared up so many points that now we perhaps know more of the life of the Salmon than of most other fishes; therefore, some account of the methods by which this splendid result has been obtained will not be out of place.

The problem of what the fish did from the time they went down to the sea as smolts until they returned as grilse was solved by a Norwegian, Herr Dahl, who adopted the plan of fishing for them with suitable nets in likely places until he found them in the open sea, following the shoals of young Herring, Smelts, Mackerel, etc.; by this same method of continued fishing throughout the year, finding out

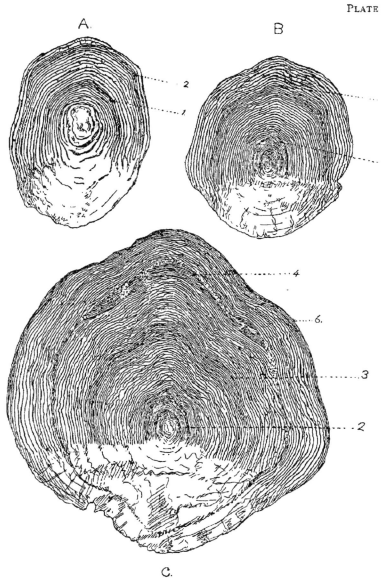

SCALES OF SMOLT (A), GRILSE (B), AND SALMON (C) (*after Hutton*)

where the fish were, and what size they were at different seasons, Dr. Dahl has established a number of facts concerning the growth and migration of Salmon and Trout.

We may next mention the marking experiments which have been carried out by the Fishery Boards in Scotland and Ireland under the respective direction of Messrs. W. L. Calderwood and E. W. L. Holt. On the Tay smolts were marked by a silver wire fastened near the base of the front of the dorsal fin, and were recaptured on their return from the sea. Kelts and clean fish also were marked in the same way, only with a numbered silver label attached to the wire, and on recapture the difference in size and weight were noted. By these methods it has been established that Salmon nearly always keep to their own river, that they do not return as grilse in the same season as they go to the sea as smolts, that some are annual spawners and others are not, and many other facts which have already been detailed.

The third means of investigating the life of the Salmon is due to the discovery by Mr H. W. Johnston of the fact that every Salmon carries its own history plainly written for those who are able to read it; this gentleman has made a detailed study of the structure of the scales as shown by the microscope, and has shown that these reveal the age and life-story of the fish. The number of scales is constant throughout life, and consequently as the fish grow the scales increase in size , the new tissue is added in the form of a series of concentric ridges, having somewhat the same appearance under the microscope as the rings of

4

growth seen in a transverse section of the stem of a woody plant, during the summer the Salmon feed more freely and grow more rapidly than in the winter, and the scales increase in size by the addition of a number of rings which are rather far apart, whilst the winter growth is represented by fewer rings closer together ; when the fish are in fresh water the scales do not grow, but their edges become more or less worn and irregular, especially at the shrinkage in girth which follows the spawning, and when the kelts reach the sea and scale growth recommences, the irregular line of the former edge is usually clearly seen and is called the spawning-mark.

In order to show the application of this method I give on Pl. V figures of the scales of a Smolt (A), a Grilse (B), and a Salmon which had spawned in the season previous to its capture (C). In all three it will be noticed that in the exposed posterior part of the scale the lines of growth have disappeared or are inconspicuous, whilst the remainder is mapped out more or less distinctly into broader zones of summer growth and narrower zones of winter growth. The smolt scale (A) is from a fish caught in May, and shows the rings formed during two complete summers and winters, and the third summer's growth just beginning. That of the grilse (B), from a 3-lb. fish taken in July, is less highly magnified, and shows the great increase caused by a summer and winter's sea-feeding (between the points marked 2 and 3) and by half a second summer in the sea (external to 3).

The third scale (C) is from a 20-lb. Salmon captured in March ; this fish spent two years in fresh water as a parr, and a few rings rather closer together than the rest at the beginning of the third summer's growth were probably formed in fresh water in April and May; then the smolt migrated to the sea and fed there for two summers and two winters, and in the following spring entered fresh water, where it stayed for a year, spawning in the autumn and getting back to the sea in the spring; then after a year in the sea it again entered fresh water and was caught, being then six years old. The scale shows the spawning-mark very clearly, and it should be particularly noticed that, although the fourth winter's growth has persisted near the anterior end, it has been completely obliterated at the sides; this is due to the great shrinkage in girth after spawning, when the upper and lower edges of the scales get frayed and worn, during the fifth summer and winter of this fish's life not only did no scale growth take place, but that of the previous winter in great part disappeared.

Having given some account of the normal life of the Salmon, and of the ways in which this has been ascertained, we may consider certain abnormalities. The earliest age at which Salmon usually become sexually mature is three and a half years, when they have been to the sea and have run up as grilse, but it sometimes happens that male parr, only 6 to 8 inches long, are full of milt, and it has been proved experimentally that they are perfectly fertile; female parr with ripe eggs are unknown, but Day mentions that of some smolts in the Howietown ponds, which were prevented from going to the sea, one developed ripe eggs in the autumn of the same season, when less than 1 lb. in weight.

This leads us to the consideration of the so-called land-locked Salmon, such as those of Lake Wenern in Sweden, which are now cut off from the sea by inaccessible falls, but which grow to a large size. In Eastern North America some of the large lakes in the New England States and in Quebec are inhabited by non-migratory Salmon, including the famous Ouananiche, which rarely grows larger than $7\frac{1}{2}$ lbs., but is a very game and active fish. We can readily understand that in large lakes with an abundant food-supply some of the Salmon might be induced to give up their journey to the sea, and if in the course of ages such lakes became inaccessible from the sea they would be inhabited by a race of land-locked Salmon.

There are no natural land-locked Salmon in the British Isles, but a smolt was sent to me from Lough Mask in June 1907 by Mr. Alick Duncan, who wrote me that the only outlets for the lake were crevices by which the water flows through underground channels to Lough Corrib, three miles distant, so that it is pretty certain Salmon do not run up to Lough Mask at the present day, and they have probably been introduced and seem to breed there. In 1881 Mr. Douglas Ogilby turned some smolts into Lough Ash in Tyrone, which has no access to the sea; two years later he captured one of them as a ripe female only $14\frac{1}{2}$ inches long, and it is possible that this lake also now possesses a stock of non-migratory Salmon.

Enough has been said to show that sea-feeding is not essential for the ripening of the sexual products, since male Salmon may become mature in the parr stage, and female Salmon, prevented from going to the sea, have been known to produce ripe ova when little more than overgrown smolts.

The Salmon has numerous enemies. Eels are especially destructive to the eggs, and Pike, Perch, Trout, and piscivorous birds prey upon the young fish during their two years of life in fresh water. Shoals of Coal-fish and other members of the Cod family are said to keep a look out off the mouths of the rivers for the descending smolts, whilst flocks of Gulls, Cormorants, etc., prey upon them in the estuaries. Seals account for large numbers of Salmon and often accompany them in their migration for considerable distances inland

Fresh-run Salmon are usually characterized by the presence of Sea-lice, little parasitic Crustaceans of the order Copepoda, which differ from the free-swimming shrimp-like members of the order in the reduction in number of segments and limbs, and in various modifications of structure which enable them to adhere to their host and to suck its blood These Sea-lice (*Lepeophtheirus salmonis*) are usually females, which measure about $\frac{3}{4}$ of an inch in length, the males being scarce and only about one-third as long. The lice are very firmly attached to the Salmon, their carapace acting as a sucking-disc and claws being dug in to increase the strength of their hold. They feed on the blood of their host, their lips forming a tubular proboscis enclosing the mandibles, which are modified into a pair of piercing spikes. The eggs are biscuit-shaped, and as they emerge from the openings of the oviducts they are cemented together by a secretion, forming a pair of egg-strings which resemble the jointed antennæ of a

lobster and may be three times as long as the
louse which carries them.

The eggs hatch out as larvæ, which swim freely
at the surface, and after a few months, in which they
change considerably, find their host and claw on to

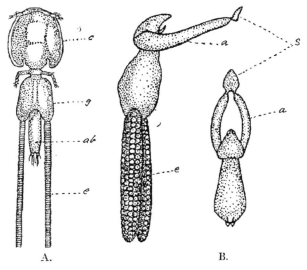

FIG. 8.—A. Sea-louse of Salmon, female, about twice the natural size
(*after Scott*).

c. carapace ; g. genital segment ; ab. abdomen ; e. egg-string.

B. Gill-maggots of Salmon, females, about four times the natural size,
from the side (*after Gadd*) and from above (*after Steenstrup and
Lütken*).

s. sucker ; a. arms ; e. egg-cases.

it ; after one or two months more they gain a new
attachment, a special frontal gland pouring out a
viscid secretion, which is pulled out into a thread
by a backward movement of the louse, and then
hardens, forming a strong and flexible connexion
which is not broken until the adult structure is

attained. The lice are unable to live in fresh water, and a few days after a Salmon has entered a river they drop off.

In fresh water Salmon may become infested with 'maggots' in their gills. These so-called. maggots (*Lernaeopoda salmonea*) are very different in appearance from the lice, and are still less shrimp-like, although they belong to the same order and resemble them in the structure of the suctorial mouth and the attachment of the larva by a frontal thread. The adult females are only one-fourth or one-third of an inch long, and a pair of appendages are modified into long arms which unite at the tip to form a sucker, which effects the attachment to the gills; the egg-cases contain several series of eggs. Males of this genus are little known, but are dwarfed, and are sometimes found attached to the females. Maggots can live in the sea, and fresh-run Salmon which have visited fresh water before usually have them in their gills.

In fresh water also Salmon may become infected with Salmon disease, which sometimes causes great mortality among them. Mr. Hume Patterson's researches, published in 1903, have established that this disease is due to a specific germ which he names *Bacillus salmonis pestis*. The bacilli invade the body of the fish when the skin has been injured or broken, and multiplying rapidly form areas of mortified flesh which are a suitable soil for the growth of the fungus called *Saprolegnia ferox*. The white patches of this fungus are usually an outward and visible sign of Salmon disease, although they do not, as was formerly supposed, constitute the disease itself. Sea water destroys the fungus, but not the bacillus, and dead fishes are sources of infection.

Those of us who look upon the Salmon as an expensive luxury, find it difficult to realize that in the eighteenth century and even later it sold at times for a penny or twopence a pound, whilst it was a regular condition in indentures of apprenticeship or in agreements between master and servant that Salmon should not be given for dinner more than so many times a week. Nevertheless, the Salmon fisheries have always been highly valued, and the fish has been the subject of special legislation and endless litigation from very early times.

Macpherson gives some interesting details showing the working of the Lakeland fisheries in the thirteenth century; at an assize held at Carlisle in 1278–79 complaint was made that the Prior of St Bees had two engines called " cupa " for catching Salmon in his pool of Stanyburn, where in times past he had but one, and the other had been set up six years ago without warrant; this resulted in an order to the sheriff to remove the second " cupa " at the Prior's expense. At the same assize a close-time was decided on, no fishing being allowed from Michaelmas to St. Andrew's Day; also, it was enacted that only nets of large mesh should be used, so that the fry could pass through At this time the kings of England used to grant rights of fisheries to great nobles, who in turn passed on a part of their privileges to the religious houses, or the latter might receive them direct from the sovereign.

At the present day the Salmon laws are probably susceptible of improvement; Mr. Robert Service, writing on Solway fishes in 1892, said: " In value

and importance Salmon far outweigh all our other
fish interests put together. But the constant flow
of law cases arising out of disputes and claims and
law-breaking in the Salmon fisheries is a real public
scandal, and unsatisfactory to everybody except the
lawyers. The late Frank Buckland would have
found a very large number of people here to agree
with him if, in his famous statement that 'more lies
have been told about the Pike than about any other
fish in the world,' he had substituted 'Salmon' for
'Pike'!" The lawyers still continue to reap their
harvest, and these remarks are as pertinent to-day
as they were nineteen years ago

On many rivers the stock of Salmon has been
kept up by the construction of ladders or passes
enabling the fish to get by insurmountable falls or
rapids, and thus gain new breeding-grounds; also,
artificial propagation is now generally resorted to,
the ripe fish being captured, stripped, as the operation
of pressing out the eggs or milt is called, and then
released; the fry are reared in the hatcheries, and
are thus protected at a period of their lives when they
are least able to take care of themselves. But when
the enormous number of smolts which fall victims
to piscivorous birds and fishes is considered, it may
be questioned whether the comparatively few turned
out by a hatchery produce much effect on the stock
of breeding fish. It seems that a Salmon river can
best be improved by the prevention of contamination
of the water, by the constructive methods already
referred to, and by the strict enforcement of legislation restricting the netsmen sufficiently to give a
fair proportion of the Salmon a reasonable chance
of reaching the breeding-grounds. The careful

treatment and liberation of captured kelts is also important, and the prohibition of the taking of parr and smolts is necessary.

None will question the pre-eminence of the Salmon among our fishes, but not every one will go so far as the Solway fishermen, who, according to Mr. Robert Service, speak only of Salmon as " Fish," and almost invariably with a peculiar deferential tone. In reply to a question as to what luck he had had, one replied, " A' had twae stanes o' fleuks, a skate, about a dizzen o' herrin', some codlin's, and *three Fish!* "

Those readers who wish to know more about the Salmon will find much of interest in the annual reports of the Fisheries Boards for Scotland and Ireland for the last ten years, inluding papers by Messrs W. L. Calderwood and E. W. L. Holt and their colleagues ; they should also consult—

(1) F. Day, *British and Irish Salmonidæ.* London, 1887.

(2) G. E. H. Barrett-Hamilton, "The Life-History of the Salmon and the Phenomena of Nuptial and Sexual Ornamentation and Development," in *Ann. Mag. Nat. Hist.* (7) ix. London, 1902.

(3) K. Dahl, " A Study on Trout and Young Salmon," in *Nyt Magazin f. Naturvidensk,* xlii. Christiania, 1905.

(4) W. L. Calderwood, *The Life of the Salmon.* London, 1907.

(5) J. A. Hutton, *Salmon Scales.* London, 1909.

(6) P. D. Malloch, *Life-History and Habits of the Salmon, Sea Trout, and other Fresh-water Fish.* London, 1910.

(7) J. A. Hutton, *Salmon Scale Examination and its Practical Utility.* London, 1910.

The Salmon of the North Pacific belong to five species, which are placed in a distinct genus (*Onco-rhynchus*); they may be mentioned here because they form the basis of the tinning industry, and because their habits are those of our Salmon in an exaggerated form; they run up in enormous numbers, the spawning fish change greatly in colour and appearance, the jaws in the males becoming excessively long and hooked at the tip, and the front teeth much enlarged; the development of the sexual organs seems to be at the expense of all the other tissues, which are so degenerated that when the act of spawning is completed every individual dies, none being in a condition to feed or get back to the sea. An un-successful attempt has been made to introduce one of these species, the Quinnat, into European waters.

CHAPTER IV

THE TROUT

Distribution — only one British species — differences from Salmon—migratory and non-migratory habit not permanent. Sea-trout : eastern and western races—food, migrations, etc.—spawning—fry—smolts—rate of growth—local names. Estuarine Trout. Freshwater Trout : variability—size—Great Lake Trout—Lochleven Trout—variation in colour dependent on environment—Trout in Monaghan—in Inverness-shire—in the Smoo—in Crassapuil—Parr-marked Trout—gillaroo—Trout of Lough Neagh—Gillaroo of Lough Melvin and Loch Mulach Corrie — malformed Trout in Yorkshire — tailless Trout of Islay—Hump-backed Trout—food and habits of freshwater Trout—Rainbow Trout

W E have already seen that although the Salmon is normally a migratory fish in our waters, yet non-migratory forms are included in the same species. This is also the case with the Trout (*Salmo trutta*), which has a somewhat different distribution from the Salmon, as it is not found in North America, and ranges from Iceland and the northern coasts of Europe southwards to the countries to the north of the Mediterranean and even to Corsica, Sardinia, and Algeria, although in these regions no Sea-trout occur. Eastward the Trout extends through Persia to the northern slopes of the Himalayas, whilst migratory Trout inhabit the Black, Caspian, and Aral Seas and their tributaries.

The Trout of Japan and of the Pacific slope of North America belong to species distinct from ours.

In the British Isles there is only one species of Trout, which is identical with the Trout of Sweden, named by Linnæus *Salmo trutta*. Linnæus regarded the Sea-trout (*S. eriox*) and the Brook-trout (*S. fario*) as distinct from the River-trout (*S. trutta*), and since his time additional names have been bestowed on a number of our native Trout, which have been recognized by various writers as different species. Some of the forms which have been thus distinguished are the Phinock or Eastern Sea-trout (*S. albus, S. brachypoma*), the Sewen or Western Sea-trout (*S. cambricus*), estuarine Trout (*S. estuarius, S. orcadensis, S. gallivensis*), the Great Lake Trout (*S. ferox*), the Lochleven Trout (*S. levenensis*), the Irish gillaroo (*S. stomachicus*), and the Black-finned Trout of some of the Welsh mountain tarns (*S. nigripinnis*).

All these are here regarded as pertaining to one variable species, which differs from the Salmon especially in the less graceful form, the deeper caudal peduncle, the stronger maxillary, and the less emarginate caudal fin; whilst the dorsal fin-rays are often fewer, the rows of scales on the tail are usually more numerous, and there are other distinctive characters, some of which have already been mentioned.

In the British Isles the Trout is found all round the coasts and in most suitable lakes and rivers, in some localities attaining a weight of as much as 50 lbs., but in small tarns and brooks often averaging only 3 or 4 ounces, and rarely growing larger. Although the silvery Sea-trout and non-migratory Brown Trout differ so much in habits and appearance, there are no structural differences, and the young are indistinguishable; other reasons for referring them to the same species may be here given.

It has been shown experimentally that Sea-

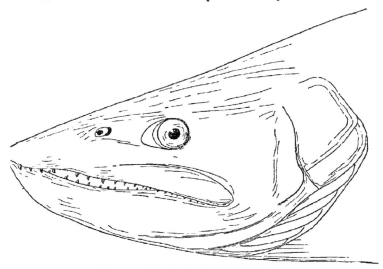

FIG. 9.—Head of a large male Trout.

trout, if prevented from going to the sea, will live and breed in fresh water; conversely Brook Trout exported to New Zealand have found their way to the sea and have given rise to an anadromous race. Estuarine Trout are often intermediate in appearance and habits between the migratory and non-migratory fish, and there is good reason to believe that in nature the ranks of the Sea-trout are reinforced by the offspring of the River-trout and

PLATE VI

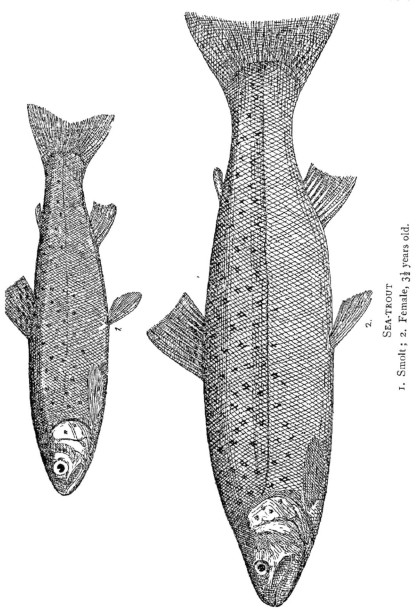

SEA-TROUT

1. Smolt; 2. Female, 3½ years old.

vice versa. The distribution of the Trout is sufficient evidence that the migratory and non-migratory fish are not distinct species, nor even races; there are no true freshwater fishes—Roach, Perch, etc.—in the Hebrides, Orkneys, or Shetlands; yet in these islands every river and loch is full of Brown Trout, which is only to be explained by the supposition that the latter have been derived from the Sea-trout, which have lost their migratory instinct in different places and at different times.

In adult Sea-trout the maxillary extends well beyond the eye, and the caudal fin, when widely spread, has the free edge straight or even convex. The coloration is silvery, with the back bluish, and with rounded or X-shaped blackish spots on the upper part of the body, usually more numerous than in the Salmon. The males differ from the females in that the whole head, but especially the snout, is longer and the jaws are stronger; in large males the lower jaw is usually hooked upward at the tip, especially in the breeding season, but never to the same extent as in old male Salmon. A comparison of the figures of the heads of old male Salmon (Pl. III) and Sea-trout (Fig. 9) and of the illustrations of Salmon parr (Pl. IV) and grilse (Pl. II, Fig 2) with those of Sea-trout (Pl. VI) will enable the reader to form a clear idea of the differences between the two species.

Two races of Sea-trout may be recognized, although in many cases it is impossible to say to which race an individual fish may belong unless one knows beforehand where it comes from. However, the Sewen (*S. cambricus*) of Wales, Ireland, and our western coasts often differ from the Sea-

5

trout (*S. albus*) of the east coast, in having a longer head, a larger mouth with stronger jaws, the suboperculum projecting backwards beyond the operculum, and the fins somewhat larger, the lobes of the caudal especially being more produced. When typical examples of the two races, of the same size and sex, are compared, these differences may be seen, but they are slight, and not always apparent.

The Sea-trout do not go so far out to sea as the Salmon, and in the spring and summer may be seen moving about in large shoals near the coasts, often leaping out of the water; in the sea they feed on Sprats, Sand-eels, shellfish, etc. They spawn in fresh water from September to January, usually in October or November, and often earlier than the Salmon in the same river. Large Sea-trout do not usually remain long in fresh water either before or after spawning, and in most rivers the main run of the breeding fish is in the late summer or early autumn.

Unlike the Salmon, the Sea-trout feed in fresh water on Minnows, worms, water-snails, shrimps, and insects, and in the larger rivers, where food is abundant, some of the breeding fish may ascend in May or June, or even before, but the Trout which form this early run are smaller than those which come later, the larger fish being less ready to forsake the stronger diet of the sea. Thus on the Tweed the June fish are said not to weigh more than 5 lbs., but those which run up in October and November weigh from 6 to 20 lbs. or more. In small rivers the autumn run is the only one; thus in the Shetlands, where there are only small burns, the Sea-trout do not enter fresh water until quite late in the autumn.

The breeding habits are very similar to those of the Salmon; the eggs hatch out in the spring and the fry remain for a time among the gravel until the yolk-sac is absorbed, when they commence active feeding, and grow rapidly during the summer. In April or May, when they are about $2\frac{1}{4}$ years old and from 4 to 8 inches long, they begin to drop down towards the sea, but unlike the Salmon smolts they are in no hurry to leave the estuaries, and indeed may grow considerably larger before doing so. Young Trout of 3 or 4 inches have already been described and compared with Salmon parr of the same size; one of these is figured on Pl. VII, Fig. 1, and we may now describe the Sea-trout smolts.

In coloration these are silvery, with the back bluish or olivaceous; there are a number of rounded blackish spots above the lateral line, also below it anteriorly and on the opercles; red spots occur chiefly on and below the lateral line; the pectoral and pelvic fins are orange, the dorsal greyish with series of blackish spots, the caudal dusky with an orange tint and with the upper and lower edges red, the anal greyish with a white anterior edge and a blackish intramarginal stripe, the adipose fin edged with brilliant orange. A figure is given (Pl. VI, Fig. 1) of a smolt of a Sea-trout from Yorkshire, a male of $7\frac{1}{2}$ inches, taken in May on its way to the sea. The maxillary nearly reaches the vertical from the posterior edge of the eye, and this should be especially noted in comparing with the Salmon smolts, in which the maxillary extends only to below the posterior edge of the pupil.

The Trout smolts, when they finally leave the estuaries, do not go so far out to sea as those of the Salmon, but feed near the coasts, and by the end of the summer are usually not much less than a foot long; in the winter they feed but little, and in the larger rivers, or where the burns flow out of accessible lochs, many of them may join the autumn runs of larger breeding fish and pass the winter in fresh water, returning to the sea in the next spring and, as a rule, running up to spawn in the following autumn. Thus we see that in the early part of the year there are migrating to the sea smolts, three-year-old fish, and larger fish which have spawned, and in the summer and autumn there run up not only breeding fish, but fish about 2½ years old, which have been only a few months in the sea.

The rate of growth is very variable, but is slower than that of the Salmon; after one summer in the sea the length is usually about 10 to 12 inches, and the weight 1 lb. or less; during the second summer's sea-feeding the length usually increases to about 18 to 20 inches, and the weight to 2 or 3 lbs.; a female fish of this size is figured (Pl. VI, Fig. 2). In the next season this weight may be doubled, so that a fish of 5 or 6 lbs. captured towards the end of the summer may generally be estimated as 4½ years old. However, some Trout appear to go farther out to sea and to grow more quickly than others, and this is especially the case with the Trout of certain rivers such as the Coquet and the Tweed; in the latter, Sea-trout more than 4 feet in length and weighing nearly 50 lbs. have been captured.

Various local names are given to the Sea-trout, and sometimes the same names are differently applied on different rivers; the smolts are often termed *Orange-fins*, and fish $2\frac{1}{2}$ years old, when entering the rivers in the summer or autumn, are sometimes called *Black-tails*, from the colour of the caudal fin, which is darker at this stage than at any other, in Devonshire these are called *Peal*. When a year older and usually about 18 to 20 inches long they are known as *White Trout* or *White-fish*, in the north of England as *Whitlings*, and in Scotland as *Herlings* or *Phinocks* (Gaelic *Fionnag*, from *fionn*, white); but all these names may also be applied to the *Black-tails*. The names *Sea-trout*, *Salmon-trout*, *Truff* (in Devonshire), *Scurf* (on the Tees), and *Sewen* (in Wales) seem applicable to any Trout which has been to the sea, whilst large fish are called *Bull-trout*, *Grey-trout*, or *Round-tails*. On the Tay the name *Bull-trout* is sometimes misapplied to the Salmon.

Sea-trout afford splendid sport to the angler, readily taking a spinning bait in the sea and rising to the fly in fresh water; as food they are usually considered a delicacy, but the colour and quality of the flesh varies enormously according to the locality.

We have already mentioned that Sea-trout smolts are in no hurry to leave the estuaries for the sea, and it is probable that a good many of them never do so, except perhaps for a few hours at a time, but become estuarine or tidal Trout; these are more or less silvery in appearance, but often retain their red spots throughout life. Some writers regard these estuarine Trout as Brown Trout acclimatized to brackish or salt water, and this may be true of many of them ; in fact, they are intermediate in appearance and habits between River-trout and Sea-trout. Fish of this type are found in the brackish waters of Loch Stennis in the Orkneys, where they attain a large size, and have been described under the name *Salmo orcadensis*.

We now come to the non-migratory freshwater Trout, and it is among these that we should most expect to find distinct forms, especially in isolated localities. However, after a careful examination of a large number of examples from all parts of the British Isles, I find a remarkable variability in those features which are easily influenced by habits or environmental conditions, but a no less remarkable constancy in those which are not so affected

The size attained differs enormously in different localities ; for example, in Lough Neagh, the largest lake in the British Isles, Trout of 30 lbs. used to be quite common, and they are said to grow to as much as 50 lbs. in weight, on the other hand, the Trout of some of the mountain streams in Wales or Cornwall average only 2 or 3 ounces, and rarely grow larger. The size is dependent to some extent on the volume of water and the amount and nature of the food available ; an important factor also is the number of fish, which depends on the area of the breeding-grounds and the extent to which enemies of the fry and young fish flourish. I have caught a good many Trout in quite a small stieam in Dorsetshire, where the usual size is from 1 to 3 lbs, but where I have never taken more than four in a day. Here I believe that so few young fish escape the perils that beset them in the form of Sticklebacks and small Pike, that those which do get through find plenty of room and plenty of food.

During growth the caudal fin changes in form from emarginate to truncate, the snout gets longer and more pointed, the eye relatively smaller, and the mouth larger. Large males also differ from the females in having a longer head, a more produced snout, and stronger jaws, the lower one often up-turned at the tip. Thus we see that in localities where the Trout grow large we may meet with fish not only bigger, but very different in appearance from any to be seen in places where only small Trout are to be found. The so-called Great Lake Trout, which has been recognized as a distinct species under the name *Salmo ferox*, is in no way different from a large Brook Trout.

The size of the maxillary bone has been said to characterize some of the races of Trout when specimens are compared with others of the same size and sex from elsewhere. However, the differences are so slight that it is difficult to appreciate them, and they may well be due to the rate of growth and the nature of the food.

The Lochleven Trout is one which has been characterized by the weakness of the maxillaries, but I find them to be much as in Trout from many other localities. Another feature said to be distinctive of the Lochleven Trout is the large number of the pyloric cæca; these appendages vary from forty-five to eighty-two in Trout from Loch-leven, more than are usually found in specimens from other British localities. However, Mr. G. Sim counted from twenty-nine to sixty-nine in fifty-two examples from Aberdeenshire, and in four from Windermere I count forty-eight to seventy-two, figures which clearly show that there is not much to be done with this character. Many authors think that the Lochleven Trout is entitled to specific rank; it has always been greatly esteemed as a sporting fish and for the table; the colour is silvery, with black spots.

The coloration of the Trout in our rivers and lakes varies enormously; some are silvery white, others quite blackish, such as the small Trout of some Welsh mountain tarns, which have received the name *Salmo nigipinnis*. The colour of the back varies from bluish grey or bright olive through different shades of green, yellow, brown, and violet to nearly black; the sides usually have silvery or golden reflections, and the hues of the back are replaced below by white, yellow, or grey. The spots, black, brown or red, stellate, round or oval, and often ocellated, differ greatly in their size, number, and distribution. Our figure (Pl. VII, Fig. 2) is of a male fish, 11 inches long, from Loch Crocach, where the Trout have the spots ocellated and large. The ground-colour has been experimentally shown to change and harmonize with the environment, especially depending on the amount of light and the colour of the bottom; the nature of the food also probably has its effect.

Dr. Gunther has written, "Trout with intense ocellated spots are generally found in clear rapid rivers and in small open Alpine pools; in the large lakes with pebbly bottom the fish are bright silvery, and the ocellated spots are mixed with or replaced by X-shaped black spots; in pools or parts of lakes

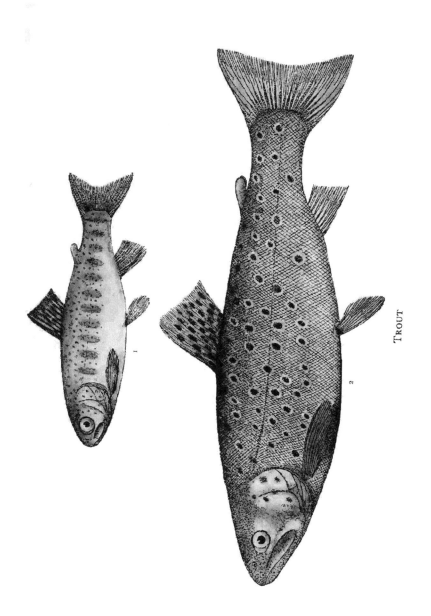

TROUT

with muddy or peaty bottom the Trout are of a darker colour generally, and when enclosed in caves or holes they may assume an almost uniform blackish coloration."

Professor Poulton has adduced some evidence that the change of colour in response to a changed environment is effected through the eye; in a clear chalky stream a few Trout were seen to be much darker than the rest, and much more conspicuous; these were found to be blind, and in consequence were ill-fed and lean.

The inconstancy of the colours of Trout is well exemplified in the following instances: Percy St. John (*Wild Sports of the West*) described a lake in Monaghan, a long irregular sheet of water of no great depth, bounded on one side by a bog, on the other by a dry and gravelly surface. On ᴇ bog side the Trout were said to be of the dark and shapeless kind peculiar to many loughs, whilst on the gravel they were of the beautiful and sprightly variety which generally inhabits rapid and sandy streams. Day gives an account of two lochs in Inverness-shire which were stocked at the same time from Loch Morar; a few years afterwards the Trout in the larger lake, which had a sandy and weedy bottom, had golden sides, were covered with numerous red spots, and had white flesh, in the smaller lake, where the water was dark coloured and the bottom rocky, they were described as having the head nearly black, the sides yellowish olive, about ten red spots and fourteen black ones present on each side, and the flesh pink.

The River Smoo in Sutherlandshire plunges through a hole in the roof of a cave and falls about

40 feet into a dark pool below; from the cave pool to the sea is a distance of only about 30 yards, but in 1876 Mr. Neil Campbell caught some Trout in this stretch and transferred them to the water above the falls, where none had previously existed. In 1882 Mr. Harvie Brown caught quite a number of these fish, which were remarkable for a line of large irregular blotches of bright crimson along the middle of the side, sometimes confluent to form an undulating band, whereas the ancestral form below the falls was said to be very ordinary looking.

Mr. Harvie Brown has described the Trout of several Sutherlandshire lochs; in Loch Crassapuil, which has a sandy bottom, they grow to a large size, and are silvery with a white belly, and the back and sides greenish with small dark spots. In Loch-na-Sgeirach, or " The Loch of the Parr-marked Trout," a catch of twenty-three trout weighed 9 lbs.; these had two rows of red spots on each side, but were especially remarkable for retaining the parr-marks, which were very distinct.

In many of the Irish lakes the fishermen distinguish with the name " Gillaroo" a Trout which differs somewhat in habits and appearance from the ordinary. The most notable things about the Gillaroo are that he is always well spotted with red, hence his name, which is derived from the old Irish *giolla*, fellow, and *ruadh*, red, and that he subsists largely on shellfish and has a remarkably hard and thick-walled stomach in consequence. The fish called " Gillaroo" in various lakes seem to agree in this, but differ somewhat in other characters, such as form, size, and value as food, according to the locality.

According to Thompson, the Lough Neagh "Gillaroo" has the upper parts yellowish with large brown spots, and towards the belly is golden, tinged with pink, and with large scarlet spots on and below the lateral line. It has a hard stomach or gizzard, is partial to a rocky bottom, and may be taken with a worm or a fly. It attains a weight of 12 lbs., and the fishermen say that it is a very inferior fish for the table. Thompson contrasts this form with the Great Lake Trout, which grows much larger and is described as silver-grey with black spots, the males having a salmon tint below, and the lower spots enclosed in orange rings; this is taken on night lines baited with Pollan or Perch, and is, according to Thompson, the common Trout of Lough Neagh.

In a case like this one would like to know whether there is an incipient species formation due to physiological isolation. Whether "once a Gillaroo always a Gillaroo" would hold good, or whether the larger Gillaroos adopt the habits and assume the livery of the *ferox*, thus taking a new lease of life. Supposing the former to be the case, it would then be a question whether the two forms keep apart when breeding, and if so, whether the offspring follow in the footsteps of their parents.

Other localities for Gillaroo are the lakes of Galway and Lough Melvin. Concerning the latter Dr. Günther tells me that a fishmonger once wrote to inform him that he had received a number of Gillaroo from Lough Melvin; he went to look at them and could feel the hard stomach by pressing with his fingers; one fine specimen he purchased, and afterwards cut it open in order to ascertain the

contents of the stomach; these proved to be a
quantity of shot packed in with bits of newspaper,
no doubt a device of the fishermen to add to the
weight, but the walls of the stomach in this
" Gillaroo" were quite thin ! ·

According to Mr. Harvie Brown, the Trout of
Loch Mulach Corrie in Sutherlandshire are " Gillaroos." Mr. Eagle Clarke tells me that these are
beautiful fish, very deep in the body, and with
brilliant red fins. In this lake also there is said to
be " a small but apparently adult form about the
size of a big Minnow and very rarely got. It may
be the young of the so-called Gillaroo, but, if so, it is
a curious departure, as it is utterly without par
bands."

Curious deformities may occur, such as an abnormally short lower jaw; one of the commonest,
found in Trout from many localities, and also in
Salmon, Pike, Perch, etc., is a shortening of the
snout so that the lower jaw projects; such fish are
termed " Bulldog - nosed Trout." Messrs. Eagle
Clarke and Roebuck (*Yorkshire Vertebrates*) write
thus on this subject : " Remarkable malformations are
observed in the Trout of Malham Tarn, and of a
beck on the western side of Penyghent. This is
manifested in the former by the deficiency of the
gill-cover in about one in every fifteen fish caught—
a calculation based upon a statement with which
Mr. Walter Morrison has furnished us of the total
number caught from 1865 to 1880. In the case of
the ' Ground Trout' of Penyghent, as they are
called, Mr. John Foster informs us that the malformation consists of a singular projection of the
under jaw beyond the upper. These aberrations are

considered to be the result of interbreeding, due to an extreme degree of isolation. The isolation of Malham Tarn is complete, it has no feeders of sufficient size for the introduction of new blood, while the overflow is absorbed by fissures in the limestone, after being swallowed up by which the water reappears—as the River Aire—after a subterranean course of two miles. The beck at Penyghent is exceedingly small, and after a short half-mile course disappears in a similar manner."

In Loch-na-Maorachan in Islay occur the so-called "tailless" Trout, in which the caudal fin-rays are abnormally short and have their ends curved together, the anal fin and sometimes also the paired fins possessing this peculiarity to a less extent. Dr. Traquair has reported on these and on similar Trout from other Scottish localities, but has only reached the conclusion that none of the causes assigned, such as wearing off of the fins against the rough bottom, nibbling by other fish, deficiency of lime in the water, etc., are adequate to produce this effect.

Hump-backed or Hog-backed Trout, with the body abnormally short and deep, occur in several localities in our islands and elsewhere; they have a malformed vertebral column, sometimes curved, usually with a number of vertebræ shortened and coalesced to a considerable extent. In 1747 Mr. Barrington pointed out that the "Hog-backed Trout of Plinlimmon" occurred only in districts where there were falls, and he advanced the theory that the eggs or fry were washed over, thus sustaining spinal injuries which led to a shortening of the vertebral column. In 1886 Day proved experimentally that concussion actually had this effect on the unhatched eggs, producing spinal curvature in the young fish; only those but slightly injured survived and grew up into Hog-backed Trout.

This theory seems scarcely applicable to Trout in ponds, especially as similarly deformed Perch are known from isolated localities in our islands and in Scandinavia; here we seem to be dealing with inherited malformations.   Mr. Harvie Brown says that Fheor's Lochan in Sutherlandshire is a small, deep, clear spring pool fed by a steep mountain burn, which glides slowly for 160 yards before entering the pool; the Trout grow to a good size and are known as " Hump-backed Trout." He caught only one specimen, which he describes as of vigorous build, very handsome, and with the hump very pronounced.

Trout are generally fond of rapid streams, but will flourish in any piece of water which is sufficiently pure, and where the Pike are not too numerous, and it is in the deeper pools or in lakes that the largest fish are to be looked for.   Trout will often lie under the shadow of a tree or of an overhanging bank, or one may be seen just behind a large stone which protects him from the full force of the current. As a rule, they feed most actively early in the morning or towards the evening, so that it has been said that the fisherman cannot be too early or too late on the water, but their appetite varies according to the nature of the day, the condition of the water, etc., which also determine whether they lie low or feed at the surface.

Their diet consists of shrimps, water-snails, insects, worms, etc., and small fish, such as Minnows or the young of their own species. Day mentions a cannibal, weighing about 1½ lbs, taken in the Tweed in June 1882, which had eleven small Trout in its stomach; the larger fish are predaceous and as destructive as the Pike On some rivers, when the Mayflies are on the water, which is usually in the early part of June, the Trout indulge in a gluttonous orgy, gorging themselves with these insects.

On most of our southern rivers the Trout are shy and crafty, and the angler who wishes to entice them into his creel has to put forward skill, patience, and cunning to outmatch them. Any one used to such fish would scarcely believe that in some waters the Trout are as bold as a Perch, or as stupid as a Pike. Yet in some of the Welsh mountain tarns the little Trout are said to dash at anything which falls on the water, so that they can be caught without any bait on the hook, whilst in the Lake District, where they grow to more than 3 feet long and attain a weight of more than 30 lbs., these large fish have been known to hold on to a bait and give play to the angler without being hooked, and after letting go, to again rush at the bait and seize it.

The usual breeding-season for non-migratory Trout is in October or November, when the estuarine Trout run up into fresh water and the lake fish either ascend the streams or seek gravelly shallows near the shores, where they form redds in much the same manner as the Salmon. The youngest spawning-fish are about two and a half years old, but it is probable that a number do not breed for the first time until they are a year older; the rate of growth for the first two years is much the same as for the Sea-trout in most localities, but afterwards must vary enormously according to circumstances.

Its sporting qualities, beautiful appearance, and delicate flavour make the Trout the most desirable of all our freshwater fishes, with the exception of its close ally, the Salmon; the latter has the advantage in size, but this is balanced by the numbers of the Trout, and its presence in all clear waters, even in quite small brooks; Salmon-fishing is for the few, but Trout-fishing is within the reach of most.

The Rainbow Trout (*Salmo irideus*) of California, which has been introduced with more or less success into some of our lakes and rivers, is a species which may usually be recognized by the spots which cover the caudal fin; the branched rays in the anal fin are often more numerous than in our species (nine to eleven instead of seven to ten), but the two forms are very closely related.

CHAPTER V

CHAR [1]

Differences from Trout : distribution—origin from migratory ancestors—long isolation in our lakes—several species in British Isles—variation in number of vertebræ—of scales—of fin-rays—of gill-rakers—coloration—size—breeding habits—extinction in certain lakes. Synopsis of British and Irish species : the Windermere Char — Scottish races of the Windermere Char—Lonsdale's Char—the Torgoch of Llanberis—the Struan of Loch Rannoch—the Haddy of Loch Killin—the Loch Roy Char—the Large-mouthed Char of Ben Hope—Malloch's Char—the Orkney Char—the Shetland Char—Cole's Char—Gray's Char—Trevelyan's Char—Scharff's Char—the Coomasaharn Char — the Blunt-snouted Irish Char — the Whiting of Lough Neagh. The American Brook Trout—the Huchen

CHAR, or fishes of the genus *Salvelinus*, differ from Salmon and Trout (*Salmo*) in that the vomerine teeth are present only as a group on the head of the bone, which is raised and has a boat-shaped depression behind it. Char may also usually be distinguished from Trout by the smaller scales and the absence of black or brown spots;

[1] I should like to place on record my indebtedness to a number of gentlemen who have sent me Char from various parts of Britain and Ireland during the last few years, and have enabled me to present an account which is much more complete than it otherwise would have been. I take this opportunity of expressing my sincere thanks to Messrs. G. Allen, J. W. Barratt ; Sir. A. J. Campbell-Orde ; Messrs.

6

the name Char is derived from the Celtic *cear,*
blood, or *ccara,* red, in allusion to the colour of the
belly.

Char are more arctic and alpine fish than Trout,
for whereas the latter range in the sea from Iceland
and the northern coasts of Europe to the Bay of
Biscay, and in the rivers to Algeria, migratory
Char are found at least as far north as 82°, and

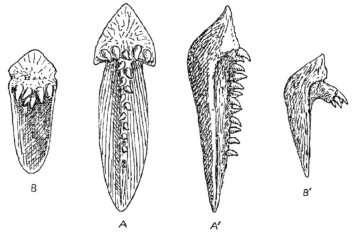

FIG. 10.—Vomer of Trout (A, A') and of Char (B, B').
A and B from below, A' and B' from the side.

southwards only to Helgeland in Norway, Iceland,
Hudson Bay, and the Kurile Islands, whilst the
non-migratory forms are principally restricted to
deep cold lakes, in our islands occurring in Scotland,

L. Dunbar, A. Duncan, G. Gatey, F. B. Henn; Capt. J. S.
Hamilton; the Earl of Lonsdale; Lord Lovat; Messrs. J. Mackay,
W. MacElligott, P. D. Malloch, W. H. Marshall, A. Matthews,
R. N. Millar, G. R. Murray, J. Murray; Col. W. H. Parkin; Dr. F.
Penrose; Mr. A. Saddler; Dr. R. Scharff; Maj. H. Trevelyan;
Messrs. J. H. Tulloch and H. M. Warrand.

the Lake District, North Wales, and Ireland, and on the continent of Europe in Scandinavia and the Alps.

As has been said, Char are chiefly found in deep cold lakes, but there are exceptions to this, and, indeed, the Char lakes of our islands show great variety. Char became extinct about one hundred years ago in Lough Neagh, which is the largest lake in the British Isles, but they still flourish in such large pieces of water as Loughs Corrib and Mask, Lochs Ericht and Rannoch, and Windermere; on the other hand, they may inhabit quite small lakes, such as one on the shoulder of Ben Hope, or the tarn at the head of Glen Roy. It often happens that they are found only in the deepest of several neighbouring lochs; for example, in Caithness they occur only in Loch Calder, which is distinguished from other lochs of that county by its depth; Loch Morar, with a maximum depth of more than 1000 feet, is said to hold Char, and Lochs Ericht and Rannoch are very deep, with greatest depths of 512 and 440 feet respectively; but depth is not essential, for in a few cases they abound in shallow lakes; a notable example is Loch Borallan, which has a maximum depth of only 21 feet.

Of Char lakes situated at an altitude of more than 1000 feet we may mention Loch Builg (1585), Loch Ericht (1153), and Loch Dungeon (1002); these may be contrasted with Lough Corrib in Ireland, Loch Fada in North Uist, or Loch Scourie in Sutherland, all of which are placed at less than 50 feet above the level of the sea.

There can be little doubt that when the temperature of Europe was lower, as during the glacial epoch, migratory Char were to be found much farther south than at the present day, and that these descended to the sea in spring, and towards the winter re-entered the rivers to spawn, as they do now in the Arctic regions. Our Char, therefore, represent a number of lacustrine colonies of one or a few migratory ancestral forms, to which they stand in much the same relation as our river and lake Trout do to the Sea-trout, with the important exception that they are isolated, and have been so since the end of the glacial epoch, perhaps nearly 100,000 years.

The word "isolated" is here used to mean that the migratory Char retreated northward, and the lacustrine colonies were thus cut off from any chance of reinforcements from the sea; the complete isolation of the Char in any particular lake or system of lakes from those of neighbouring lakes or lake systems may have taken place either long ago or quite recently, according to local geological changes.

At the present day our lake and river Trout are continually reinforced from the ranks of the Sea-trout, and when they are found in localities now inaccessible to the latter there does not seem to have been any long-standing isolation; consequently there are in our islands no races or incipient species of Trout which are recognizable and definable.

The case is quite otherwise with the Char, which I have studied with particular attention for some years. The Char of the Lake District (*Salvelinus willughbii*) and its close ally, *S. colii*, from the west of Ireland, probably come nearest to the migratory ancestral form. In several of the Scottish lakes the Char differ so little from those of the Lake District that they may be regarded as races of *S. willughbii*. But in a number of other lakes they present such remarkable peculiarities, due to their long-continued isolation, that they can only be looked upon as distinct species.

I am quite aware that some authors contend that there is only one species of Char in our islands, whilst some would not even recognize the various forms as distinct races. Certainly our species of Char are recent species and geographical species; they are of quite another nature from widely distributed forms such as the Pike or Roach, which have probably persisted unchanged during the whole of the time that the evolution of the *Salvelini* has proceeded. Nevertheless, they differ from each other in characters which are used to define species in other groups, and which may, therefore, be regarded as specific.

In fishes numerical characters are often of great importance for the distinction of species, especially as they are not subject to change during growth. I have counted the vertebræ in various British and Irish Char with the following result :—

Locality	Number of Vertebræ
Llanberis, North Wales .	61, 61
Windermere .	60, 61, 62
Coniston . . .	61
Loch Grannoch, Kirkcudbrightshire	62, 63, 64
,, Doon, Ayrshire . .	62, 62
,, Rannoch, Perthshire .	62
,, Builg, Banffshire .	62
,, Killin, Inverness-shire .	62
,, Bruiach ,, .	60, 62
,, Borollan, Sutherlandshire . .	60

Locality	Number of Vertebræ
Loch Baden, Sutherlandshire . . 61	
„ Loyal, „ . . 59, 60, 61, 61, 61, 61	
„ Scourie „ . . 64, 64	
„ Stack „ . . 64	
„ under Ben Hope, Sutherlandshue . 63, 64, 64	
„ Hellyal, Orkneys . . . 59	
„ Girlsta, Shetlands . . . 62	
„ Fada, North Uist . . . 62	
Lough Luggala, Wicklow. . . 62, 62, 63	
„ Mask, Galway . . . 62	
„ Eske, Donegal . . . 62, 63	
„ Melvin, Fermanagh . . 58, 59, 59, 60, 60	

These numbers are interesting in that they show a differentiation of the Sutherlandshire Char into two groups, those with fifty-nine to sixty-one vertebræ in the lochs of the southern and eastern parts of the county and those with sixty-three or sixty-four vertebræ inhabiting the lochs of the north-west; whilst in Ireland the Char of Lough Melvin seems to be well separated from the rest by the fewer vertebræ (fifty-eight to sixty instead of sixty-two or sixty-three).

The number of scales in a longitudinal series, *i.e.* the number of oblique rows of scales running downwards and backwards to the lateral line, is subject to very considerable variation when a number of specimens from the same locality are compared; thus in thirty-eight Char from Loch Loyal I count from 126 to 178 scales in a longitudinal series, and these I cannot regard as specifically distinct from the Char of the Lake District, in which I count 160 to 200 scales. However, the number of scales is always important, and in some of our Char is a very valuable specific character. The single example known to me of the Char of Lough Owel in Westmeath has 186 scales in a longitudinal series, and if I had been in doubt as to whether its other peculiarities warranted me in describing it as a new species, the fact that I had examined sixty Char from other parts of Ireland, none of which had more than 168 scales, would have had considerable weight. The species most notable for the large size of the scales is the Lough Melvin Char; in twenty-six examples of this form I count 128 to 162 scales, in striking contrast to the small-scaled Char of Loch Killin, which has 180 to 220.

The number of fin-rays is of little value in this group; in the Loch Loyal Char the dorsal fin has from eight to eleven branched rays, and these figures cover the variation in all the British or Irish forms. In Ireland the Char of Wicklow and Kerry usually have a somewhat longer anal fin than those from other localities, the branched rays numbering from eight to eleven instead of from seven to nine.

The number of gill-rakers on the lower part of the anterior branchial arch varies from eleven to sixteen in Char from Loch Loyal and in those from Windermere. These figures also cover the entire variation in the Irish Char, except the form inhabiting Lough Coomasaharn in Kerry, which has the gill-rakers longer and more numerous than the rest, numbering eighteen or nineteen on the lower limb of the anterior branchial arch.

In coloration the Char vary as much as the Trout; the back is usually bluish grey or bluish black, but may be lilac, lead-coloured, greenish, brownish, or bright olive; this hue descends on to

the sides and shades below into a silvery white or
into an orange or crimson of a greater or less
intensity, according to the locality, sex, and season ;
pink, orange, or red spots are usually, but not always,
present, sometimes only below the lateral line,
sometimes covering the back and sides and even
extending on to the dorsal and caudal fins, which
are usually greyish or blackish ; scattered white spots
may also be present on the sides ; the caudal fin
is often tinged with red, sometimes uniformly,
sometimes near the edge only, or with a reddish
blotch below or one on each lobe ; in the most
intensely coloured forms the lower fins are red with
the anterior edge white, but the pectoral usually
has a dark greenish shade near the white edge,
and the red colour of the pelvics and anal may
sometimes be greatly reduced.

In some lakes the Char grow to a good size ; in
Windermere they reach a weight of nearly 3 lbs,
and Mr. Malloch tells me that he has seen one
of 2½ lbs. from Loch Ericht. The size attained
evidently depends chiefly on the food-supply and
the number of fish, and does not correspond with
the area of the lake, Char grow to 2 lbs. in Loch
Killin, which is quite a small loch, whereas in the
much larger Loch Doon they never weigh more
than a few ounces. In lakes where the Char run
small, such as Haweswater and Loch Doon,
they feed very constantly at the surface, and are
often caught with an artificial fly, but where molluscs,
shrimps, etc., are plentiful the Char usually feed at
the bottom, and frequenting the deep water are
seldom seen.

The breeding season is from September to March,

the actual time of spawning varying greatly in different lakes, presumably being affected by the temperature of the water and other circumstances. The Char spawn on beds of gravel in shallow or moderately deep water, either in the lakes or in tributary streams, forming redds like Salmon and Trout. In some places, for example Loch Killin and Loch Grannoch, they suddenly come on to the shallows, and within a few days all have spawned and gone back into the deep water, and they exhibit in this an extraordinary regularity, appearing year after year at the same time almost to a day. The habits of the Char in Windermere are markedly different, for their spawning time is protracted, extending from November to February or even March.

Char have certainly died out in several lakes during the last hundred years, but the causes which have led to their extinction are not well ascertained. Mr. F. B. Henn tells me that since the introduction of Trout into Lough Gortyglass the Char are less abundant than formerly, and this leads one to suspect that the stocking of Hellyal Lake in the Orkneys with Trout may have had something to do with the disappearance of the Char. For although Char and Trout may get on perfectly well together in a lake which they have inhabited for thousands of years, it does not follow that Char used to having a lake to themselves will be able to compete successfully with a suddenly introduced stock of Trout. In Lough Neagh overfishing, in Lochleven the reduction of the area of the lake, in Ullswater pollution of the river in which they used to spawn, have been assigned as possible causes of the extinction of Char. However this may be, it is certain that if Char were to die out in the next few thousand years at the same rate as they have done in the last century they would remain only in a very few lakes in the British Isles, and if, for example, these were Girlsta, Killin, Llanberis, and Melvin, all zoologists would agree that our Char were four well-marked species. The extinction of annectant or intermediate types can hardly be said to lead to the formation of species, but it certainly leads to their recognition.

Some of the characters used in the distinction of the species will appear in the following :—

Synopsis of the British and Irish Species of Char

I. BRITISH SPECIES

A. Body moderately elongate, the greatest depth rarely more than one-fourth or less than one-fifth of the length (to the base of the caudal fin).

1. Adult males [1] with the snout conical or subconical, more or less acute, and with the lower jaw pointed anteriorly, rarely slightly shorter than the upper and sometimes projecting beyond it.

a. Interorbital region more or less convex, except in young specimens; teeth moderate.

a. Maxillary never extending far beyond the eye.

[1] Young specimens and females usually have the snout blunter and the lower jaw weaker.

Lower jaw two-thirds the length of head in males of 10 to 12 inches, less in smaller examples; snout as long as the diameter of eye in males of 6 or 7 inches . . . . 1. *willughbii*

Lower jaw more than two-thirds the length of head, and snout longer than the diameter of eye in males of 7 inches . . . . 2. *lonsdalii*

        β. Maxillary extending well beyond the eye in males of 8 to 11 inches . 3. *maxillaris*

    b. Interorbital region quite flat; teeth rather strong; snout somewhat produced and acutely conical . 4. *perisii*

2. Adult males with the snout sub-conical, obtuse, and with the lower jaw obtusely pointed anteriorly and a little shorter than the upper . . . . . 5. *mallochi*

3. Adult males with the snout obtuse, with the upper profile decurved, and with the lower jaw rounded anteriorly when seen from below, shorter than and included within the upper.

    a. Maxillary strong, extending to below the posterior edge of the eye in males of 8 inches; eye moderate; fins large . . . 6. *killinensis*

    b. Maxillary short, not reaching the vertical from the posterior edge of the eye in males of 8 inches

Eye small, its diameter less than one-fifth the length
of head; fins small . . . . 7. *inframundus*
Eye large, its diameter not much less than one-
fourth the length of head; fins rather large
8. *struanensis*
   B. Body very elongate, the greatest depth about
      one-sixth of the length . 9. *gracillimus*

II. IRISH SPECIES

   A. Snout conical or subconical; lower jaw
      pointed anteriorly, not included within
      the upper; anal fin with 7 to 9 branched
      rays.
        1. Not more than 170 scales in a
          longitudinal series; maxillary ex-
          tending a little beyond the eye in
          males of 11 or 12 inches.
          a. 12 to 16 gill-rakers on the
            lower part of the anterior
            branchial arch.
            a. Snout subconical, de-
              curved; teeth feeble or
              moderate.

Depth of body 4 to 5 in the length; least depth of
caudal peduncle about two-fifths the length of
head; pectoral fin extending from one-half to
nearly three-quarters of the distance from its base
to the pelvics; 138 to 168 scales in a longitudinal
series; 62 or 63 vertebræ . . . . 10. *colii*
Depth of body 3½ to 4 in the length; least depth
of caudal peduncle one-half or nearly one-half
the length of head; pectoral fin extending two-
thirds to nine-tenths of the distance from its

base to the pelvics; 128 to 162 scales in a
longitudinal series , 58 to 60 vertebræ . 11. *grayi*
β. Males with the snout
produced, acutely
conical, and the
teeth rather strong
12. *trevelyani*
  b. 18 or 19 gill-rakers on the
  lower part of the anterior branchial arch . 13. *fimbriatus*
 2. More than 180 scales in a
 longitudinal series; maxillary extending to or a little beyond the
 posterior edge of pupil in a male
 of 11 inches, snout conical, rather
 short . . . . . 14. *scharffi*
B. Snout obtuse, with the upper profile decurved ; lower jaw rounded anteriorly,
shorter than and included within the
upper, interorbital region flat ; anal fin
with 8 to 11 branched rays . 15. *obtusus*

WILLOUGHBY'S or the WINDERMERE CHAR (*Salvelinus willughbii*) inhabits Windermere and several
other lakes in the Lake District, namely, Coniston
Water, Goat's Water, Wast Water, Ennerdale Water,
Buttermere, Crummock Water, and Lowes Water.
All these are deep lakes, the deepest being Wast
Water, which has a depth of 258 feet, they vary in
size from Goat's Water, which is a small tarn,
to Windermere, the largest lake in England and,
studded with islets, famous for the beauty of its
scenery.

The present writer has examined Char from Windermere, Coniston, Buttermere, and Crummock Water, and presumes that those of the other lakes mentioned belong to the same species.

In this form, described by Dr. Gunther in 1862, there are eight to ten branched rays in the dorsal fin, and seven to nine in the anai, the scales in a longitudinal series vary from 160 to 200, *i.e.* that is the number of oblique rows of scales descending to the lateral line; there are from eleven to sixteen moderately long gill-rakers on the lower part of the anterior branchial arch, and the vertebræ number sixty to sixty-two.

The adult fish is usually 10 to 12 inches, rarely as much as 18 inches, long, and has nearly the form of a Trout, the depth measuring from one-fourth to one-fifth of the length to the base of the caudal fin; the snout is subconical, with the jaws equal anteriorly, the mouth slightly oblique, with the maxillary extending to or a little beyond the vertical from the posterior edge of the eye, and the teeth moderately strong. The diameter of the eye measures about one-sixth, the width of the convex interorbital region one-third, or a little less than one-third, of the length of the head. The dorsal fin is moderately elevated, the longest ray measuring one-half to two-thirds the length of the head, and the pectoral fin, when laid back, extends from one-half to three-fourths of the distance from its base to the insertion of the pelvic fins. The females have a somewhat smaller head, blunter snout, shorter paired fins, and duller coloration than the males, whilst young specimens are distinguished from the adult fish just described by their blunter snout, smaller mouth,

PLATE VIII

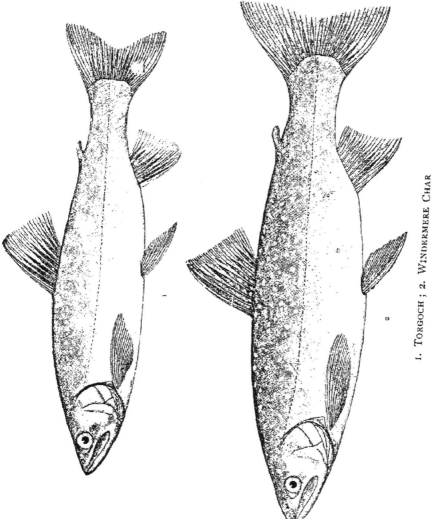

1. TORGOCH ; 2. WINDERMERE CHAR

larger eye, narrower and flatter interorbital region, broader opercles, and smaller fins.

The following description of the coloration of four specimens from Crummock Water, each nearly a foot long, was drawn up when they reached the writer, in splendid condition, in the month of October: "Back and sides bluish, with silvery reflections and with numerous pink spots everywhere; faint traces of nine to twelve parr-marks; belly red; snout, upper part of head, and sometimes the maxillary blackish; cheeks and opercles silvery, with shades of blue, green, or pink; lower jaw, branchiostegals and thoracic region white; iris golden, pupil black; dorsal and caudal fins blackish, with or without pale pink spots at the base; pectoral dusky, tinged with red, sometimes with the upper ray whitish; pelvics and anal similar, but redder and with strongly marked white anterior edges."

Specimens from Windermere and Coniston taken earlier in the season (April and May) were silvery, with the back bluish or bright olive, and the fins pale, but towards the autumn the Char of Windermere is coloured like those described above from Crummock Water.

In Windermere the Char inhabit the deeper parts of the lake, but in the summer months, especially towards the evening, they swim at the surface. During the season, from March to September, they are netted in considerable numbers, and are either eaten fresh or used for "potting." They may sometimes be captured by means of a spinning Minnow, but the most deadly way of spinning is from a boat by means of a plumb-line, a long and strong line with a heavy lead sinker, with shorter lines bearing the 'Minnows' attached at intervals.

These 'Minnows' are by no means elaborate, as they are cut out of sheets of metal by the fishermen, and are furnished with but a single triangle of hooks attached to the tail end.   In plumb-line fishing two stout rods are used, one projecting from each side of the boat, which is rowed slowly along, the Char being permitted to hook themselves and to signify the same by ringing a bell, which is placed at the end of the rod.   This ingenious device for getting the fish to ring his own knell was, I believe, invented by the Chinese, and I was not aware that it was in use in England until I paid a visit to Windermere.   On this lake the Char but rarely take a fly, and of late years are said to feed much less at the surface than formerly, perhaps owing to the increased number of petrol-driven launches on the lake.   At Windermere I found the impression prevalent that there were two species of Char in the lake; the reasons for this belief are the differences in colour and in time of spawning.   The variations in colour are undoubtedly due mainly to age, sex, and season, and the so-called 'Silver Char' are either young fish, or if adult usually females taken early in the season.

The breeding season extends from November to February or March, and I was informed that the majority of the fish spawn either in November or in February, that the early fish are smaller and more silvery, and make their redds either in the River Brathay or in the shallow water near the edge of the lake, whereas the February fish are larger and redder, and spawn in comparatively deep water. It would seem that most of the fish which are spawning for the first time do so in November, but there is room for further investigation, which, however, is not at all likely to establish the existence of two species or races of Char in one lake.

In most of the lakes inhabited by the Windermere Char their capture is a profitable industry. In Goat's Water they are said to be plentiful but very small, averaging eight to the pound, and to rise to the fly readily. The Ennerdale Char spawn in the so-called 'Char Dub,' a long pool in the River Liza, about three hundred yards above the lake; in an account of a visit to this place in November, 1850, the Char were described as blackening the bottom of the pool. "Many thousands certainly were there, and in a proper light the gleam and twinkle of their multitudinous white-edged fins was a pretty and singular spectacle."

The Irish naturalist, Thompson, wrote: "When at the inn at Waterhead, at the northern extremity of Coniston Water, during a tour to the English lakes in June, 1835, a number of Char from this lake were kept alive by our host in a capacious wooden box or trough, into which a constant stream of water poured. They were fine examples of the species, about a foot in length. Here I was informed that a supply of this delicate fish was always kept up, that the 'curious' visitor might gratify his taste at any season by having fresh Char set before him at the rate of ten shillings for the dozen of fish." Some years before, Pennant wrote of Coniston: "The fish of this water are Char and Pike; a few years ago the first were

7

sold for 3s. 6d. per dozen, but thanks to the luxury of the times are now raised to eight or nine shillings."

That potting Char is not a modern idea is shown by Defoe, who in 1769 wrote of the Char-fish of Winander Mere: " It is a curious fish, and, as a dainty, is potted and sent far and near by way of present." In the seventeenth century Char-pies seem to have been a favourite luxury, and the price of Char was only 3s. a dozen, as appears from the account-book of Sir Daniel Fleming of Rydal, which includes such items as: "Mar. 23, 1662; for the carryage of a charr-pie unto my aunt Dudley at London, at 2d. per lb., £0, 6s. 0d." and "June 29, 1665; for twelve charrs when Mr. Dugdale was here . . . 3s. 0d." [1]

Our figure (Pl. VIII, Fig. 2) is of an adult male fish, nearly a foot long, from Crummock Water. A young male from Windermere is also figured (Pl. IX, Fig. 2), and shows well the features enumerated above as characteristic of immature fish.

As has already been stated, a number of the Scottish lakes are inhabited by Char which cannot be regarded as specifically distinct from those of the Lake District, and these must be considered next.

In the south-west of Scotland there are three lakes which contain Char, namely, Lochs Grannoch and Dungeon in Kirkcudbrightshire and Loch Doon in Ayrshire; the statement that Char occur

[1] The quotation from Defoe and the items from Fleming's account-book are taken from Macpherson's interesting account in the *Fauna of Lakeland*.

in Loch Inch in Wigtownshire is incorrect, and
due to confusion with Loch Insh in Inverness-shire.
Of these three lochs the largest is Loch Doon,
6 miles long and 2 to 6 furlongs broad, whilst
Loch Dungeon is quite small, three-fourths of a
mile long and one-fourth of a mile broad.

The Loch Grannoch Char is a small race of
S. *willughbii*; the specimens I have seen measure
from 7 to 9 inches in length, and they are adult
fish taken when spawning. These have nearly
the form and proportions of Windermere Char
10 to 12 inches long, *i.e.* they assume the adult
characters at a smaller size than in Windermere;
some of the males have the external sexual characters very pronounced, the lower jaw projecting
slightly, and the pectoral fins nearly reaching the
pelvics. The coloration of the male fish is violet-
black above, brilliant orange below, with round
scarlet spots on the sides; the dorsal and caudal
fins are dusky, and the lower fins reddish with
more or less distinct white anterior edges: the
females are similarly but less brilliantly coloured,
the orange on the belly being sometimes reduced
to a mere tinge.

These Char keep in the deeper parts of the lake,
except for about ten days in the middle of October,
when they spawn on the shallows.

The Char of Loch Dungeon is known to me only
from a single male specimen 6 inches long, a
plump fish rather noticeable for the weakness of the
lower jaw and the breadth of the opercular bones
when compared with the Grannoch Char. Mr.
Robert Service regards this form as nearer to the
Grannoch fish than to those of Loch Doon, from

which lake he has sent to the British Museum a number of specimens, 6 or 7 inches long, captured with a fly in the month of May.   These are silvery fish with a bluish back, rather slender, and very similar to young Windermere Char taken early in the season, but usually with the paired fins longer, the mouth larger, and the opercles narrower.

In the highlands of Scotland, Lochs Builg in Banffshire, Bruiach in Inverness-shire, Morie in Ross-shire, Borollan, Loyal, and Baden in Sutherlandshire, and Calder in Caithness contain Char which have their peculiarities, but which I am not at present inclined to separate specifically from *S. willughbii*, to which also the Char of Loch Fada in North Uist is very closely allied.

In some of these the scales average larger than in the Char of the Lake District, and the vertebræ are also subject to some variation according to the locality.   The North Uist form seems to lack spots entirely, the bluish-black hue of the back shading into the uniform reddish colour of the body below the lateral line.

Ullswater and Haweswater belong to the system of the River Eden, and lie somewhat apart from the lakes inhabited by the Windermere Char, from which they are separated by Helvellyn and Shap Fell.   The Ullswater Char is believed to be now extinct, owing to the pollution of the stream in which they spawned by refuse from the mines, but Char are still to be found in Haweswater, although not very abundantly.

LONSDALE'S   or   the   HAWESWATER   CHAR (*Salvelinus lonsdalii*) is a much smaller fish than

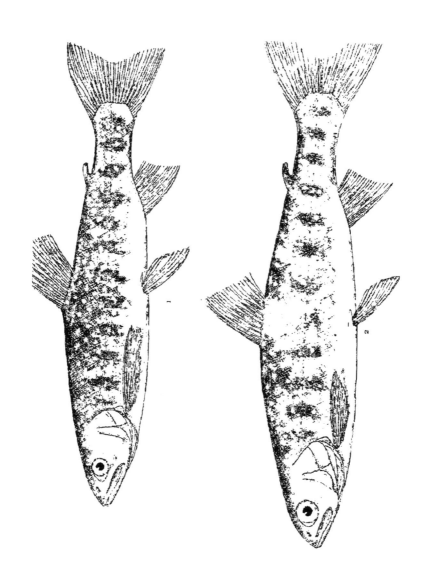

that of Windermere, reaching a length of only 7 inches, and a weight of about 3 ounces. Colonel W. H. Parkin, to whom I am indebted for the types of this species, which I named after the Earl of Lonsdale, informs me that the Char caught in Haweswater hardly vary at all in size. Compared with the Windermere Char this species is seen to differ in the longer and more pointed snout, smaller eye, narrower opercular bones, and larger fins. Especially noteworthy, however, is the great length of the lower jaw, which is much longer than in Windermere Char of the same size, and, compared with the length of the head, is even longer than in large adult examples of the latter species. The colour is bluish black above, orange below, with numerous orange spots on the sides. The specimen figured (Pl. IX, Fig. 1) is a male 7 inches long, and for comparison a somewhat larger young male Windermere Char has been put on the same Plate. This Char is said to feed mainly on insects and to take the artificial fly freely.

THE TORGOCH (*Salvelinus perisii*) is found in three mountain lakes at or near Llanberis in Carnarvonshire, and also in Lake Corsygedol in Merionethshire. The Welsh name "Torgoch" signifies "Red Belly," and the specific name *perisii* is taken from Llyn Peris, one of the lakes inhabited by this fish. The Torgoch was first properly described by Dr. Gunther in 1862, and is a rather small fish, growing to a length of 9 or 10 inches, and averaging 4 to 6 ounces in weight. The head is long, with the snout somewhat produced and pointed, the mouth large and oblique, the lower jaw

long and in the males usually projecting, and the
teeth rather strong; the interorbital region is quite
flat, and narrower than in the Windermere Char.
The example figured (Pl. VIII, Fig. 1) is a male,
9½ inches long, from Llanberis.

In the summer months the Torgochs appear in
shoals on the surface of the water, their dorsal fins
just showing; they are then occasionally taken with
a fly, but the best sport is obtained in September
and October, fishing with a worm, either from a
boat or from the shore; in this way as much as
45 lbs. has fallen to one rod in a day.

At Llanberis the lower lake, Lyn Padarn, is deep
and has a rocky bottom, and towards the end of
November the Char pass up the small stream which
connects the two lakes in order to spawn in the
shallows at the lower part of the upper lake, Lyn
Peris. It was said by Pennant that the Char of
Llanberis had all been destroyed by water from the
copper mines, but he was evidently mistaken.

THE STRUAN CHAR (*Salvelinus struanensis*), of
Loch Rannoch in Perthshire, is known to me from
fine examples, 7 to 8½ inches long, three males and
two females, the largest of these, a male, is figured
on Pl. X, Fig. 1. This species differs from those
previously considered in the short blunt snout and
inferior mouth; the rounded lower jaw is shorter
than the upper and included within it when the
mouth is closed; the maxillary is short and broad,
only reaching the vertical from the posterior edge of
the pupil and measuring but little more than one-
third of the length of the head; the teeth are feeble;
the eye is large, its diameter measuring one-fourth

or a little less than one-fourth of the length of the head, and the interorbital region is narrow and flat, not much wider than the diameter of the eye.

The rather long head (length one-fourth or more than one-fourth of the length of the fish) and the elevated dorsal fin, the longest ray of which measures three-fourths the length of the head in the males and a little less in the female, are also notable features. The scales in a longitudinal series number 158 to 180, and there are eight or nine branched rays in the dorsal fin, whereas the Killin Char, the next to be described, usually has ten or eleven.

This species was first described in 1881 by Sir John Gibson-Maitland, who netted over a score of them one day in September of that year, when they were about to spawn. According to his account their coloration is beautiful: above the lateral line, black shot with metallic blue, and below it, claret coloured, shaded with steel blue; spots, salmon colour, indistinct, about twenty above the lateral line and sixty on or below it; parr-marks visible; dorsal fin without spots, with a black marginal band; caudal with a broad dark marginal band; anal, greyish, darker below, with a white anterior edge; pectoral, claret coloured.

All the examples captured had their stomachs immensely distended with water-fleas (*Daphnia pulex*); their average length was 8½ inches, and the average weight 3 ounces.

From the structure of the mouth one would suppose that the Struan Char is a bottom feeder; the size of the eye also suggests that this species lives at considerable depths, and we know that for the greater part of the 9 miles of its length Loch Rannoch is a very deep lake, from 350 to more than 500 feet deep.

Other lakes belonging to the Tay system are said to contain Char, which may be similar to the Struan. Char were at one time pretty plentiful in Loch Leven, but have long been extinct, the last being captured in 1837; it is noteworthy that this extinction followed an extensive drainage operation in 1830, which reduced the area of the lake to three-fourths of its original dimensions, and decreased its depth by $4\frac{1}{2}$ feet.

THE HADDY (*Salvelinus killinensis*), of Loch Killin in Inverness-shire, is one of the most distinct species of Char in the British Isles. Loch Killin is quite a small lake, and its greatest depth is only 67 feet; nevertheless, its Char grow to a considerable size, attaining a length of 16 inches. I have figured a male of only 8 inches (Pl. X, Fig. 2), for better comparison with a Struan of nearly the same size.

In the blunt snout and the subterminal mouth, with the lower jaw shorter than the upper, this fish resembles the Struan, but differs from that species in the stronger maxillary, the smaller size of the eye, and the greater width of the interorbital region. It has a rather clumsy form, as the head is large and obtuse, and the body deep and but little compressed. The fins are large, especially in the males, and the scales are smaller than in most British Char, numbering 180 to 220 in a longitudinal series. The coloration is rather sombre, the back and sides olivaceous or plumbeous, the belly silvery or yellowish.

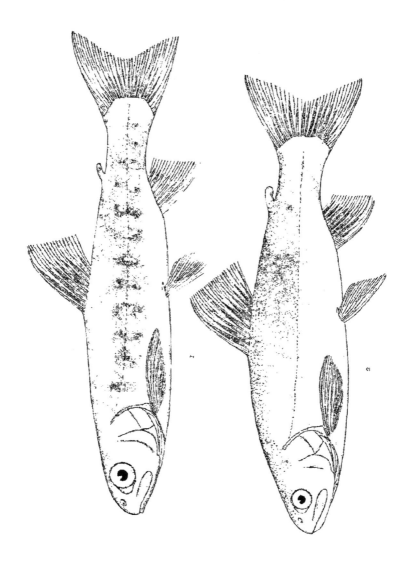

In some large specimens examined the stomach contained pieces of weed, insect larvæ, and small bivalve shells, but in a young fish the stomach was filled with small crustaceans—*Entomostraca.* For the greater part of the year the Killin Char live in the deepest part of the lake, and are never seen; but at a certain season—for about ten days in the middle of September—they come on to the shallows to spawn.

The following interesting account of this species, and of the Char of Loch Corr, was given by the Irish zoologist, Thompson, in 1840: " The Loch Corr specimen—a ' Northern Char '—is in beauty of colour and elegance, combined with strength of form, the finest example I have seen; it is of a fine deep grey on the upper parts, becoming lighter towards and below the lateral line, about which it is adorned with white spots; on the lower portion of the sides it is silvery, and beneath of the most brilliant red. This specimen is 16 inches in length, and, with another of similar size, was taken by my friend when angling with an artificial fly, on the 25th of September. The other, which was eaten, was excellent and high-flavoured, the flesh firm and red. Loch Corr is described to me as a deep mountain-lake or basin, less than a mile in length, with rocks rising precipitously above it at one part; at another it is shallow and sandy, and here this fish is taken in some quantity when spawning. A beautiful clear river issues from the lake. About 15 miles from Loch Corr is Loch Killin, situated in the pastoral vale of Stratherrick. Three specimens of Char have thence been brought me. They are remarkably different from the Loch Corr example, are of a clumsy form, have very large fins like the Welsh fish, and are very dull in colour—of a blackish leaden hue throughout the greater part of the sides, the lower portion of which is of a dull yellow, no red appearing anywhere. So different, indeed, is this fish from the Char of the neighbouring localities, that it is believed by the people resident about Loch Killin to be a species peculiar to this lake, and hence bears another name—'Haddy' being, strangely enough, the one bestowed upon it. This fish is only taken when spawning, but then in great quantities, either with nets, or a number of fish-hooks tied together, with their points directed different ways. These, unbaited, are drawn through the water where the fish are congregated in such numbers that they are brought up impaled on the hooks. The largest of my specimens is 16 inches in length, and others of similar size were brought to my friend at the same time, when about a 'cart-load' of them was taken. The flesh of some was white and soft. They contained ova the size of peas. At this very time the Char from the neighbouring Loch Corr were in high condition. This is one out of numerous instances which might be adduced respecting the different period of spawning in con-tiguous localities."

The Char of Loch Roy in Inverness-shire is known only from a single specimen 6 inches long, captured by Mr. Cholmondeley Pennell in 1862, and presented by him to the British Museum. It is in every way similar to the Killin Char, except that the head is perhaps a little smaller and the scales are larger, numbering only 160, instead of 180 to 220 in a longitudinal series.

This resemblance to the Killin Char is interesting;
the two lakes are about 15 miles distant from each
other, and separated by high mountains, but both
are south of the Glenmore chain of lochs, into which
the streams which issue from them flow, that from
Loch Killin into Loch Ness, and the River Roy into
Loch Lochy. Loch Roy is described by Mr. Pennell
as a small mountain tarn situated directly below an
almost perpendicular cliff, he caught the Char with a
fly in the month of July during a violent snowstorm.

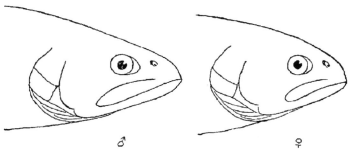

FIG. 11 —Heads of male (♂) and female (♀) of Large-mouthed Char from
Ben Hope.

THE LARGE-MOUTHED CHAR (*Salvelinus maxillaris*), of a small isolated loch under Ben Hope in
Sutherlandshire, is a very distinct species, to which
I have given the name *S. maxillaris*, on account of
the notable length of the maxillary, which extends
back far beyond the eye in adult males; the head is
longer and the interorbital region narrower than in
the Windermere Char. As will be seen from the
accompanying figures, the sexual differences are well
marked.

The colour of the back and sides, with the dorsal
and caudal fins, dark plumbeous; belly, brilliant

orange, small orange spots on the sides, mostly below the lateral line; pectorals, greenish with a red margin; pelvics and anal, reddish, with a white anterior edge; caudal with an orange margin

Eleven examples sent to me by Mr. John Murray measure 8 to 11 inches in total length; from these I have selected for illustration (Pl. XI, Fig. 1) a male of 10 inches. Three female fish about 8 inches long, from Loch Stack, are very similar to the Ben Hope Char, but until I have seen male specimens from this locality I cannot be certain that the fish of the two lochs are identical.

MALLOCH'S CHAR (*Salvelinus mallochi*) is another Sutherlandshire form which occurs in Loch Scourie, where four examples, 8 to 12 inches long, were captured by Mr. P. D. Malloch, the well-known naturalist of Perth, in whose honour I named the species. The largest of these, a male, is figured on Pl. XI, Fig. 2.

This is a rather short-headed, blunt-snouted, and small-mouthed Char, with small scales (188 to 200 in a longitudinal series) The lower jaw is obtusely pointed anteriorly and slightly shorter than the upper even in adult males; the paired fins are short. The back and sides are slate coloured, covered with numerous pale spots, the belly whitish, tinged with orange

As has been mentioned above, the Char of Lochs Scourie and Stack and of the loch under Ben Hope have a larger number of vertebræ (63 or 64) than those of the rest of Sutherlandshire (59 to 61).

THE ORKNEY CHAR (*Salvelinus inframundus*) is a well-defined species, which has the interest that it

PLATE XI

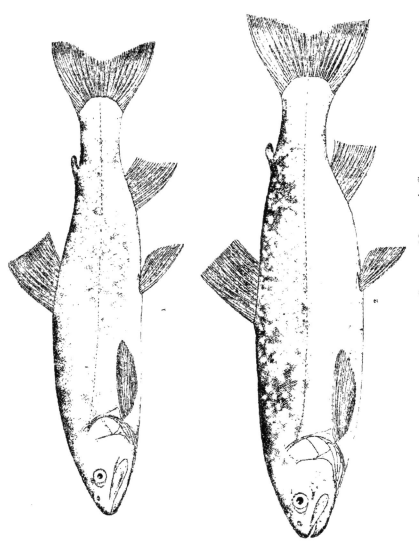

1. LARGE-MOUTHED CHAR; 2. MALLOCH'S CHAR

is probably extinct. It used to inhabit the Loch of Hellyal in the Island of Hoy, and in 1862 Dr. Trail captured two of them, fishing with a worm. These are both males, 7 to 8 inches long, and are now in the British Museum; one of them is shown on Pl. XII, Fig. 2.

In the rather slender body, the blunt snout, and the rounded lower jaw included within the upper, this species resembles the Rannoch Char; but there the resemblance ceases, for the head is short, the eye small, the interorbital region convex and rather broad, and the fins small; moreover, the caudal peduncle is more slender than in the Struan, twice as long as deep.

During the last few years Mr. William Cowan has very kindly made several attempts on my behalf to get more examples of this interesting fish, but without success. Mr. P. Middlemore, who owns the lake, has also made unsuccessful efforts to catch some Char; none have been captured since he has been the proprietor, and he believes they are extinct. Char are not known from any other lakes in the Orkneys.

THE SHETLAND CHAR (*Salvelinus gracillimus*) is also a very distinct form. It is found in the Loch of Girlsta, Tingwall, whence some have been sent to me by Mr. J. S. Tulloch, who tells me that Girlsta is the only Char loch in the Shetlands. This species has the body more elongate than any other Char, the greatest depth being contained from five and a half to six and a half times in the length of the fish, measured to the base of the caudal fin; the form of the snout, blunt and somewhat truncated, is also peculiar ; the lower jaw is not, or scarcely, shorter than the upper, the eye is rather large, and the fins are well developed. The colour of the back and sides, with the dorsal and caudal fins, is bluish grey, that of the belly silvery or orange; orange spots are present on the sides.

This is a small species, attaining a length of 8 inches; the male example figured (Pl. XII, Fig. 1) measures an inch less.

COLE'S CHAR (*Salvelinus colii*), originally described by Dr. Gunther in 1863, from Lough Eask in Donegal, is now known to inhabit the neighbouring Lough Derg, several lakes in Connaught, Lough Gortyglass in Clare, and Lough Currane in Kerry. In Donegal and in the small lakes of Connemara it grows to a length of about 8 inches, but in Loughs Conn, Mask, Corrib, and Gortyglass specimens a foot long are to be met with.

Cole's Char is almost identical with the Windermere Char, and were it not that it exhibits very constant characters in widely separated localities in Ireland, it might well be placed as a race of *S. willughbii*. It has the body moderately elongate (the greatest depth contained four to five times in the length of the fish, without the caudal fin) ; the caudal peduncle rather slender, its least depth about two-fifths the length of the head ; the snout sub-conical, decurved, with the jaws equal anteriorly ; the mouth moderately large, the maxillary extending a little beyond the eye in males of 12 inches, the interorbital region a little convex and broad (its width one-third the length of the head in the adult fish); the pectoral fins not very long, extending

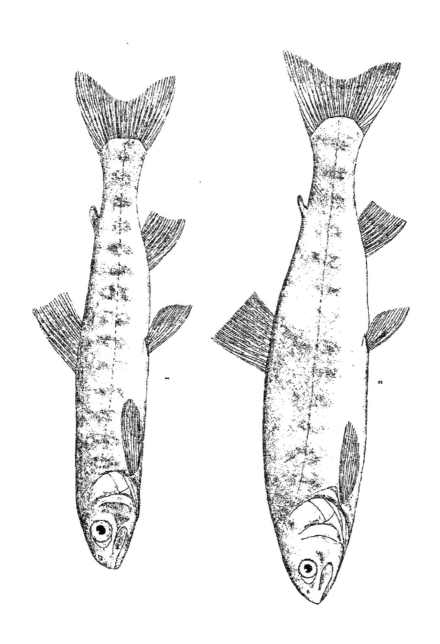

from one-half to nearly three-fourths of the distance from their base to the origin of the pelvic fins; and the scales comparatively large, 138 to 168 in a longitudinal series.

The coloration of examples from Lough Eask has been described as bluish black above, silvery or orange below, with orange spots on the sides, but I have been told by Mr. A. Matthews that the Connemara fish are greenish, with the spots on the sides white.

In Lough Eask the Char spawn in November, and at that time were formerly taken in huge quantities by the country people.

GRAY'S CHAR (*Salvelinus grayi*), the so-called "Freshwater Herring" of Lough Melvin in Fermanagh, originally described by Dr. Günther in 1862, has an average length of 10 to 12 inches, and differs from Cole's Char in having the body usually more compressed and deeper (depth $3\frac{1}{3}$ to 4 inches in the length), the caudal peduncle shorter and deeper (its least depth nearly one-half the length of the head), the pectoral fins usually longer (extending two-thirds to nine-tenths of the distance from their base to the pelvic fins), the scales sometimes a little larger (128 to 162 in a longitudinal series), and the vertebræ fewer (58 to 60 instead of 62 or 63); the example figured (Pl. XIII) is a male, 10 inches long.

In most of the species of Char the females are distinguished from the males by the less brilliant coloration, the smaller head, blunter snout, shorter maxillary, and weaker lower jaw, and by the lesser development of the fins; in some of the British forms these sexual differences are very marked. In the Lough Melvin Char, however, it is almost impossible to distinguish the sexes from external characters, as was recognized so long ago as 1841 by Thompson, from whose account of this species we extract the following :—

"To the kind attention of Viscount Cole I am indebted for twelve char from Lough Melvin, sent immediately after capture. In the accompanying note, dated November 15, his lordship remarked, ' I can procure you any number you wish, as the people are now taking them in cartloads ; the flesh of such as I send is white and soft, and different from what that of char is in any other lough.'

"These specimens, which are in a fresh state and excellent condition for examination, are all from 10 to 12 inches in length, and differ greatly from those of Loch Grannoch in presenting little or no beauty of appearance. The upper half of the body, in both sexes, is of a dull blackish lead colour, unrelieved by spotting in any but three or four individuals, which exhibit a considerable number of minute spots which are merely of a paler shade than the surrounding parts, and consequently inconspicuous ; for more than half the space between the lateral line and ventral profile they are of a dull lead colour, without any spots except in the individuals just noticed ; the lower portion of the sides varies in individuals from a pale to a rich salmon colour, which latter is seen in only one or two examples. The dorsal fins are of a uniform grey and transparent ; in some, when closely examined, there appear roundish spots of a paler colour ; pectorals dusky grey, darker towards the tips, except at the lower portion, which,

Plate XIII

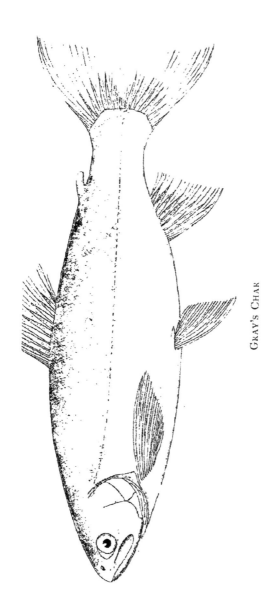

Gray's Char

partaking of the colour of the part of the body on which it rests, is of a pale pinkish white; ventrals in the brighter-coloured individuals with a white marginal line; in the duller-coloured examples this does not appear, but all have the two or three first rays and their connecting membiane dusky, and the remainder red and of a deeper hue than on any part of the body; anal fin partaking at the base of the colour of the part of the body to which it is attached, dusky towards the tip; white margin to the first ray in some of the brighter-coloured specimens only; caudal fin grey, of different shades in all, in the brightest individual varied with red, which appears at the base of the lower lobe.

" The males are generally more gracefully formed than the females, and most of them rather brighter in colour, but there is no external character so strikingly different as to lead to a certain knowledge of the sex; some of the largest-finned are females— in the Loch Grannoch char the males had much the larger fins and the sex was as unerringly distinguished by the colour as by the form, the accuracy of the distinction in both cases being established by dissection.

" The milt and roe were in these specimens ready for exclusion. The remains of food were found in only one out of the twelve specimens, and appeared to be a portion of the case of a caddis worm. The vertebræ, as reckoned in two specimens, male and female, were sixty in number.

" Lord Cole informs me that this fish is called ' Freshwater Herring' at Lough Melvin, although in the same part of the country the term ' char' is applied to the more ordinary state of the species as

8

taken in other lakes. Its differing from the so-called char, in being an insipid bad fish for the table and pale in the flesh, is the chief reason of its being considered distinct from it."

TREVELYAN'S CHAR (*Salvelinus trevelyani*) was described from a single male specimen, 8 inches in length, figured on Pl. XIV, Fig. 1. This was sent to me in 1906 from Lough Finn in Donegal, by Major H. Trevelyan. It is distinguished from Cole's Char by the longer head, narrower interorbital region, produced pointed snout, and strong dentition.

I have recently received a second smaller example, a female, from Captain J. S. Hamilton, which proves that in this form the sexes differ considerably, for, having a smaller head, a shorter and blunter snout and weaker teeth than the male fish, this specimen is quite as similar externally to a male Cole's Char as to the male of its own species.

Of the British species the Haweswater Char (*S. lonsdalii*) shows considerable resemblance to this one, but has the snout shorter, the teeth smaller, the fins larger, etc. Char have been recorded, but not described, from a number of loughs in northern Donegal, and it may be that *Salvelinus trevelyani* is the species which inhabits them.

SCHARFF'S CHAR (*Salvelinus scharffi*), from Lough Owel in Westmeath, was recently described by me from one example, nearly a foot long, in the collection of the Dublin Museum, lent to me by Dr. R. Scharff. This species has the broad head and short snout of Cole's Char, from which it differs in that the snout is more acutely conical, the mouth smaller and more oblique, etc., and in the smaller scales, which number 186 in a longitudinal series. It is a silvery fish, with the back bluish grey.

In 1851 Thompson was informed that Char were common in Lough Owel. Mr. Black, a gamekeeper, had seen sixty to seventy dozen taken in a draught-net in the summer, and in June he had seen two dozen taken in a day with the natural and artificial fly, particularly the former, the "green drake" being

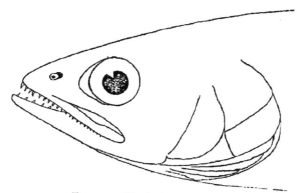

FIG. 12.—Head of Scharff's Char.

the favourite; in these cases the fly was sunk three or four feet beneath the surface. For a few successive years not a Char would be taken in the lake, and then they would again appear to be as numerous as ever. Mr. Black described them as "very round in the body," and this applies well to the specimen now in the Dublin Museum; he gave the usual weight as $1\frac{1}{4}$ to 2 lbs., but thought he had seen some of 3 lbs.

Char have been recorded from Loughs Ennel and Belvidere in Westmeath, Eaghish in Monaghan,

Drumlane in Cavan, and Loughnabrach in Long-
ford, none of which are very far from Lough
Owel.

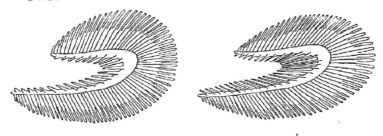

FIG. 13.—Anterior branchial arch of Cole's Char (*a*) and Coomasaharn Char
(*b*), showing the difference in length and number of the gill-rakers.

THE COOMASAHARN CHAR (*Salvelinus fimbriatus*)
takes its specific name (*fimbriatus*, fringed) from the
long and numerous gill-rakers.   In other Irish Char
eleven to sixteen may be counted on the lower limb
of the anterior branchial arch, but in this species
eighteen or nineteen.   This form is known only
from a single female example, 6 inches long, from
Lough Coomasaharn in Kerry; it is very similar to

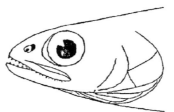

FIG. 14.—Head of Coomasaharn Char.

Cole's Char, but has the interorbital region flatter
and narrower, its width only a little more than one-
fourth of the length of the head.

THE BLUNT-SNOUTED IRISH CHAR (*Salvelinus obtusus*) is known from Loughs Luggala and Dan in Wicklow, and Killarney and Acoose in Kerry. It is distinguished from other Irish Char by the short, blunt snout and the rounded lower jaw, which is included within the upper when the mouth is closed ; in most other characters it agrees with Cole's Char, but has the interorbital region narrower and flatter, its width being contained three and one-third (adult) to three and three-fourth times (young) in the length of the head, whilst the anal fin is usually longer.

This species attains a length of 8 inches, and a male specimen of that size from Killarney is figured (Pl. XIV, Fig. 2). It is not unlike the Struan (*S. struanensis*) of Loch Rannoch ; on comparison I find that it differs especially in the shorter head, always less than one-fourth of the length of the fish and the narrower but longer maxillary, which in males of 8 inches reaches the vertical from the posterior edge of the eye, and measures two-fifths the length of the head ; also the scales are usually larger, numbering 142 to 166 in a longitudinal series, and the dorsal fin is lower, the longest ray measuring two-thirds the length of the head or less.

Char are said to occur, or to have occurred, in several lakes in Kerry, Cork, and Waterford, which will probably prove in some cases to contain *Salvelinus obtusus*. Smith, in his *History of Waterford*, quoted by Thompson, wrote: " In these mountains [Cummeragh] are four considerable loughs, two of which are called by the Irish Cummeloughs and the other two Stilloges, the largest of which contains about five or six acres. In these loughs are several

kinds of trout; and in the former is a species of fish
called charrs, about 2 feet long—the male grey, the
female yellow bellied; when boiled the flesh of these
charrs is as red and curdy as a salmon, and eats
more delicious than any trout." In 1839 Thompson
was informed that in the lakes at the source of the
River Lee in Cork, formerly celebrated for their fine
Char, which were abundant, these could not be pro-
cured, and had become nearly if not quite extinct,
their destruction being attributed by people in the
neighbourhood to the increased numbers of Pike.

One other Char, certainly extinct, completes our
account of the Irish fishes of this group, and this is
the so-called "Whiting" of Lough Neagh, described
and figured in 1812 in Dubourdieu's *History of the
County of Antrim*. There can be no question that
this fish was a Char, as its appearance, habits,
season for capture, etc, were well described. In
1837 Thompson offered a handsome reward for a
Lough Neagh Whiting, but in vain. He was told
that none had been taken for ten years, but that
ten years before that they were abundant; as
many as five hundred had been taken in a draught
of the net, and this not in the breeding season. A
deep part of the lake, 36 fathoms, was called the
Whiting-hole, from being the chief haunt of the
species. The causes of its extinction were unknown,
although overfishing had been suggested

To conclude the account of the Char, mention
must be made of the so-called Brook Trout (*Salvelinus fontinalis*) of Canada and the Northern United
States, which has been introduced into some English
streams. This species is easily recognized by its
coloration, the back, with the dorsal and caudal

PLATE XIV

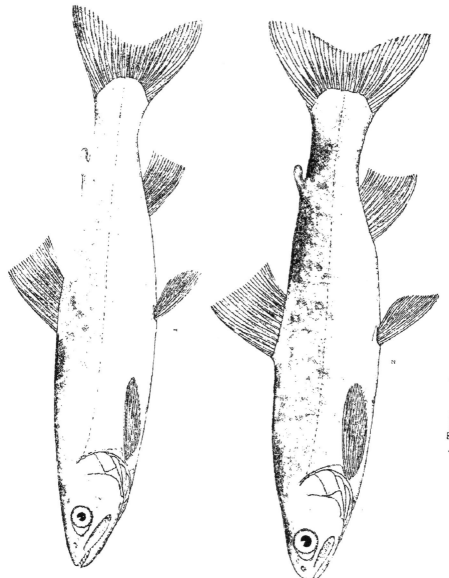

1. Trevelyan's Char : 2. Blunt-snouted Irish Char

fins, being mottled or barred with black or dark olive.

The Huchen (*Hucho hucho*) of the Danube is a large predaceous fish, silvery and black spotted like the Salmon, but structurally more nearly related to the Char. The vomer has a raised head, with a hollowed out depression behind it, but the vomerine teeth form a regular transverse series, connecting those of the palatines, whereas in the Char they are arranged more or less definitely in the form of a V. It is to be hoped that attempts to introduce this species into the Thames will prove unsuccessful.

CHAPTER VI

WHITEFISH AND THE GRAYLING

Whitefish. The British species : the Lochmaben Vendace—
the Cumberland Vendace—the Lough Neagh Pollan—the
Lough Erne Pollan—the Shannon Pollan—the Powan—the
Schelly—the Gwyniad—the Houting—the Grayling

THE Whitefish, or Salmonoid fishes of the
genus *Coregonus*, differ from Salmon, Trout,
and Char in their larger scales, and in having the
mouth small, with the teeth very minute or absent ;
they are very like Herrings in general appearance,
but the presence of an adipose dorsal fin indicates
their pertinence to the Salmon family.

The Whitefish resemble the Char in their distribution, inhabiting the northern parts of North
America, Europe, and Asia, and on the continent
of Europe ranging southwards to the alpine lakes
of Switzerland and the Tyrol; most of the rather
numerous species are freshwater fishes, but in arctic
regions they are marine, entering the rivers for
breeding purposes, and some of the more southern
forms also still retain this migratory habit.

Whitefish are seldom caught by the angler in
Europe, although some of the American species are
said to rise to the fly; however, they are valued as
food, and are netted in all localities where they
occur in sufficient number.

The British and Irish forms of Whitefish may be grouped as follows :—

Mouth
- terminal, with the
  - lower jaw projecting—*Vendaces*
  - jaws equal in front—*Pollan*
- subterminal or inferior, with the snout
  - vertically or somewhat obliquely truncated —*Powan*, *Schelly*, and *Gwyniad*
  - produced, conical—*Houting*

THE VENDACE OF LOCHMABEN (*Coregonus vandesius*) apparently derives its name from the old French word " Vendese," the modern equivalent of which is " Vandoise," the Dace.

The pretty little town of Lochmaben in Dumfriesshire has no less than seven lochs in its immediate vicinity, and is actually built on the banks of three of these—the Castle, Kirk, and Mill Lochs, the first and last of which are inhabited by the Vendace The Castle Loch, so named from the ruins of Bruce's Castle on its banks, is by far the largest, but it is of no great size, and one can easily walk right round it in an hour; its greatest length is three-fourths of a mile, and its breadth nearly as much; this loch is not more than about 20 feet deep, whereas the much smaller Mill Loch (3 furlongs long and $1\frac{1}{4}$ broad) is said to have a depth of 70 feet.

The Vendace is a small fish, reaching a length of only 9 inches, and in form and coloration shows considerable resemblance to a Herring. The body is fusiform, but compressed; the head is pointed, with the mouth oblique, the maxillary extending to below the anterior part of the eye, and the lower jaw projecting; the branchial arches are furnished with numerous long and slender gill-rakers.

In these features the Vendace agrees with some

migratory species which ascend the Siberian rivers from the Arctic Ocean and with several continental forms inhabiting the countries round the Baltic; but it differs from all of these in the larger scales, which number only from sixty-two to seventy-three in a longitudinal series and six or seven between the lateral line and the base of the pelvic fin, and in the more posterior position of the dorsal fin, the origin of which is nearly equidistant from the end of the snout and the base of the caudal fin, and is not or but little in advance of the pelvic fins; the rather deep body and the large fins are also notable characteristics of this species, which is usually a silvery white fish with the back greenish blue; sometimes the dark colour of the back extends down to the lateral line, below which the body may be of a golden tint.

The Lochmaben Vendace is gregarious, swimming in shoals; its food consists mainly of minute shrimp-like organisms—*Entomostraca*—which abound in the lakes, and which may occasionally be supplemented by small insects, larvæ, etc. The shoals retire to the deeper parts of the lakes in warm weather, and at the end of the summer approach the shores to spawn, which they do usually in November, when they are frequently seen at the surface of the water.

They are captured by means of sweep-nets with a small mesh, and it has been observed that in the summer they are most successfully taken during a dull day and in a sharp breeze, when they forsake the depths and swim near the edges of the lochs in a direction contrary to the wind. Formerly a club, consisting of twenty or thirty of the neighbouring gentry, used to meet annually in July to enjoy the

Plate XV

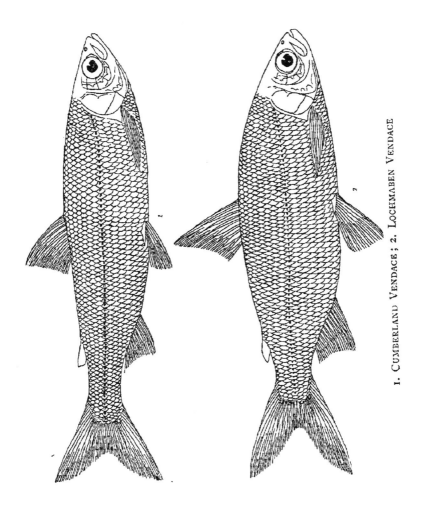

1. Cumberland Vendace; 2. Lochmaben Vendace

sport of fishing, and afterwards feasted on the catch.

According to Mr. Robert Service, a second club, the St. Magdalene's Vendace Club, was on somewhat democratic lines ; occasionally as many as two thousand people assembled at the fishing, and various athletic sports were engaged in after the netting was over for the day.

Both Vendace clubs became defunct, but about four years ago one was re-established, but so far has not had a very successful career, the annual netting procuring only about half a dozen Vendace on each occasion. When I was at Lochmaben, it was explained to me that this probably did not indicate that the fish were getting scarce, but was due to the fact that the fishermen had not had the good fortune to strike a shoal.

The Vendace is a great delicacy, resembling the Smelt in flavour, but its reputation as a food-fish is enhanced by its restricted distribution and the difficulty of procuring it.

The inhabitants of Lochmaben look upon the Vendace as a mysterious fish peculiar to their lakes, a fish in which they take a pride, and one concerning which there are many curious traditions.

The legend that the Vendace was introduced from the Continent by Mary, Queen of Scots, who visited Lochmaben in 1565, is still believed in by a number of people. The improbability that in those days so delicate a fish could have been brought so great a distance and successfully established in strange waters, and the fact that the fish is unknown on the Continent, although a closely related form inhabits Derwentwater, sufficiently dispose of this curious tale, the origin of which can only be understood by a visit to this part of Scotland, which may almost be regarded as dedicated to Queen Mary, so numerous are the castles where she stayed, the hills by which she rode, and the ports whence she embarked, according to local traditions.

The idea once prevalent that if a Vendace was taken from the water it would die, and its immediate return prove of no avail, was disproved by keeping some alive for an hour or two in a bucket of water, and I was told that some had even been transported and turned alive into Lochleven, but that nothing more had been seen of them. The belief that these fish were able to exist without food was due no doubt to the fact that they were never seen feeding, and could not be caught by bait.

The lochs at Lochmaben contain a number of other species of fish, including Perch, Roach, Chub, Bream, and Pike; the last named is said to be very destructive to the Vendace. The right of fishing in all the lochs was granted to the burgh of Lochmaben by a charter of James VI, and this privilege is still enjoyed by the townspeople.

THE CUMBERLAND VENDACE (*Coregonus gracilior*) inhabits Derwentwater and Bassenthwaite lakes; it is of the same size as and is closely related to the Lochmaben species, but as a rule has a more slender body, a shorter head, and smaller fins, whilst the dorsal fin has usually more rays. The number of scales in a longitudinal series (sixty to seventy-two) and of branched rays in the anal fin (nine to twelve) is the same in both forms. The

differences between the two may be tabulated
thus—

LOCHMABEN VENDACE

(*Coregonus vandesius*)

The depth of the body is
contained three and two-thirds
to four and one-fourth times
in the length of the fish, the
length of the head four and
one-third to four and two-
third times.

The caudal peduncle is once
to once and a half as long as
deep.

The longest dorsal ray is as
long as or a little shorter than
the head, and the pectoral fin
extends considerably more
than one-half of the distance
from its base to the pelvic fins.

The dorsal fin has seven or
eight, rarely nine, branched
rays.

CUMBERLAND VENDACE

(*Coregonus gracilior*)

The depth of the body is
contained four to five times in
the length of the fish, the length
of the head four and one-half
to five times.

The caudal peduncle is once
and a half to twice as long as
deep.

The longest dorsal ray is
shorter than the head, and the
pectoral fin does not usually
extend much more than one-
half of the distance from its
base to the pelvic fins.

The dorsal fin has nine,
sometimes eight or ten,
branched rays.[1]

These differences are illustrated on Plate XV,
where examples of the two forms, in each case about
6 inches long, are figured.

At one time Vendaces were plentiful enough in
Derwentwater and Bassenthwaite; thus in 1856
Davy wrote that during the preceding eight years
a good many had been netted in both lakes.

At the present day they are generally considered
to be scarce, but Mr. H. A. Beadle, a keen local

[1] I have counted the dorsal fin-rays in quite a number of Vendace,
and the result is as follows. Of 30 Lochmaben Vendace, 6 had 7, 21
had 8, and 3 had 9 branched rays in the dorsal fin, and of 14 Derwentwater Vendace 3 had 8, 9 had 9, and 2 had 10 branched rays in the
dorsal fin.

angler and naturalist, told me recently that he had
seen shoals of fish swimming near the surface in
Derwentwater, which he believed to be Vendace.
Certainly they are never caught by anglers, but the
net fishermen do not try for them, and so far as I
could gather from conversation with them they
thought that to catch Vendace they would have to
use a net of smaller mesh and fish farther from the
shore than they are allowed to do at present.   It was
curious to come from Lochmaben, where the Cumberland Vendace had never been heard of, and to talk
with the Derwentwater boatmen, who knew not of the
Lochmaben Vendace, but were well acquainted with
their own species, although they never see it except
on the rare occasions when one is found, dead or
dying, either floating on the water or washed ashore.

THE POLLAN (*Coregonus pollan*) of Lough Neagh
probably gets its name from the same source as the
Pollack, *i.e.* the Celtic word *Pollag*, Whiting.   It
differs from the Vendaces in several respects; the
mouth is less oblique and the lower jaw does not
project beyond the upper; the scales are smaller,
numbering seventy-four to eighty-six in a longitudinal series, and seven and a half to nine between
the lateral line and the base of the pelvic fins; the
dorsal fin is placed a little farther forward, and the
pelvic fins are inserted in or a little in advance of
the vertical from the middle of its base.   There are
nine to eleven branched rays in the dorsal fin and
eight to eleven in the anal; the fins are comparatively small, the longest dorsal ray measuring
about two-thirds the length of the head, the pectoral
extending about one-half of the distance from its

PLATE XVI

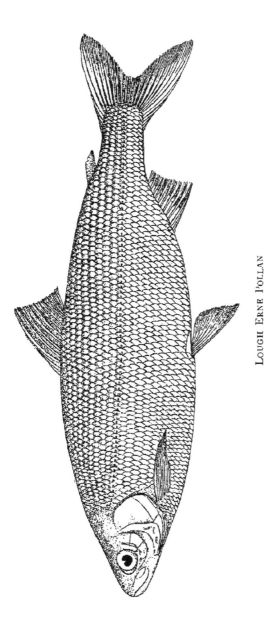

LOUGH ERNE POLLAN

base to the pelvics. The coloration is silvery, with the back bluish, and the fins more or less blackish.

The Pollan is not very different from an arctic marine species, which enters the rivers of Siberia; but there are no European Whitefish at all closely related to it. The Pollan usually average 9 or 10 inches in length, and specimens of this size weigh about 6 ounces. The Irish naturalist, Thompson, mentioned one of $2\frac{1}{2}$ lbs., the largest he had heard of.

The staple food of the Pollan appears to be a minute Entomostracan (*Mysis relicta*), but examination of their stomachs has shown that they also appreciate insect larvæ, shrimps, small bivalves, and the fry of other fishes; they have sometimes been taken with the artificial fly.

The Pollan spawn in November and December, choosing a place where the bottom of the lake is hard or rocky; during the spring, summer, and autumn they approach the shore in large shoals, and are netted in considerable numbers. Thompson records that in the first week of September, 1834, more than seventeen thousand were taken on one day by three or four draughts of the net, and were all sold on the spot at the rate of 3s. 4d. a hundred (123 fishes). At the present day they are exported to England, and are esteemed a well-flavoured fish when fresh, but they do not keep well. It may be mentioned that Lough Neagh is the largest lake in the British Isles, and that Lough Erne, which also contains Pollan, almost rivals it in size.

THE LOUGH ERNE POLLAN (*Coregonus altior*) differs from the Pollan of Lough Neagh chiefly in that it is usually a deeper fish, and often has the

9

scales more numerous in a transverse series.　The
characters in which the two forms are not identical
may be compared thus—

LOUGH NEAGH POLLAN	LOUGH ERNE POLLAN
(*Coregonus pollan*)	(*Coregonus altior*)
Depth of body contained three and three-fourths to four and one-half times in the length of the fish ; caudal peduncle once and one-half to twice and one-fourth as long as deep. Maxillary extending to below anterior one-third of eye or beyond. Seven and one-half to nine scales between lateral line and base of pelvic fin, nineteen to twenty-two round the caudal peduncle.	Depth of body contained three and one-third to four times in the length of the fish ; caudal peduncle once and one-fourth to once and two-thirds as long as deep. Maxillary extending to below anterior one-fourth or one-third of eye Eight and one-half to ten scales between lateral line and base of pelvic fin, twenty-one to twenty-four round the caudal peduncle.

This form is not so abundant as the Lough Neagh
Pollan, but still occurs in sufficient numbers to be
occasionally netted for the market, fetching 8d.
per lb. in Belfast and 10d. per lb. in England ; in
size it averages somewhat larger than the Neagh
fish ; the example figured (Pl. XVI), nearly a foot
long, is one of a fine series of Lough Erne Pollan
kindly sent to me by Major H. Trevelyan.

THE SHANNON POLLAN (*Coregonus elegans*)
inhabits the lakes of the Shannon system, especially
Loughs Ree and Derg.　It has the elongate body
of the Neagh Pollan combined with the shorter
caudal peduncle of the Erne fish ; the mouth is
smaller than in either, the maxillary extending only
a little beyond the vertical from the anterior margin
of the eye ; the scales number seventy-eight to

ninety-two in a longitudinal series, usually nine or ten, sometimes eight, between the lateral line and the base of the pelvic fin, and twenty-two to twenty-six round the caudal peduncle.

Previous to the drainage of the Shannon in 1845–46 this fish was fairly abundant, but it is now rare; one from Lough Ree, 15 inches long, is said to be the largest taken for years. Thompson says the Shannon fishermen called the fish a "Cunn," and believed that it migrated to the sea, merely conjecturing this because they caught them in the eel-nets at the time that the Eels were descending.

Pollan have been reputed to inhabit Lough Corrib, but if so they are very scarce; one said to have come from that lake, on rather doubtful authority, is in every way similar to those from the Shannon. In 1852 Mr. Ffennell is said to have exhibited Pollan from Lough Neagh and from Killarney to the Dublin Natural History Society, and to have directed the attention of the meeting to the difference in shape of the head and of the gill-covers in the specimens from the two localities. This is the only evidence forthcoming as to the occurrence of Pollan in Killarney; recently the lake has been tried by Mr. Holt with suitable nets, but without any positive result, so that it seems possible that in this case Shad may have been mistaken for Pollan.

THE POWAN (*Coregonus clupeoides*) inhabits Loch Lomond, one of the largest and most picturesque lakes in Scotland, and also the neighbouring Loch Eck; its native name seems to be a variation of the word "pollan," and it is also known as the Freshwater Herring, this latter title justifying the specific name *clupeoides* given to it by the French naturalist Lacepède in 1803.

In form and coloration the Powan is very similar to the Pollan of Lough Neagh, from which it differs especially in having the mouth smaller and sub-terminal or inferior, with the maxillary just reaching the vertical from the anterior edge of the eye, and with the lower jaw included within the upper. The snout is vertically or somewhat obliquely truncated, and the fins are larger than in the Pollan, the longest dorsal ray measuring from four-fifths to seven-eighths of the length of the head, and the pectoral extending more than one-half of the distance from its base to the pelvics.

The usual length of the Powan in Loch Lomond is about a foot, but it grows to 18 inches or more, sometimes attaining a weight of 2 lbs.; in Loch Eck it is smaller, and three examples from that locality sent to me by Professor D'Arcy Thompson are only 8 or 9 inches long; one of these is figured (Pl. XVII, Fig. 1).

Parnell's account of this species, written in 1838, is as follows: "These fish are found in Loch Lomond in great numbers, where they are named Powan or Freshwater Herring. They are caught from the month of March until September with large drag-nets, and occasional instances have occurred in which a few have been taken with a small artificial fly; a minnow or bait they have never been known to touch. Early in the morning and late in the evening large shoals of them are observed approaching the shores in search of food, and rippling the surface of the water with their fins as they proceed. In this respect they resemble in their habits the Lochmaben Vendace and the common salt-water Herring. They are never seen under any circumstances in the middle of the day. From the estimation these fish are held in by the neighbouring inhabitants they are seldom sent far before they meet with a ready sale, and are entirely unknown in the markets of Glasgow. In the months of August and September they are in the best condition for the table, when they are considered well-flavoured, wholesome, and delicate food. They shed their spawn from October to December, and remain out of condition until March."

In 1891 Mr. A. Brown published some additional details; the Powan were still very numerous, he describes them as having a strong odour like the Smelt, so that in the summer and autumn, when they float in large shoals at the surface, the surrounding air for a distance is tainted with their scent. Owing to their sluggishness and their habit of lying near the surface they are a favourite food of gulls and cormorants. They feed on entomostraca, insects, worms, and weeds, and spawn in the shallow bays on stones or gravel; in the middle of the summer, fry about 2 inches long may be seen in the creeks.

THE SCHELLY (*Coregonus stigmaticus*) is said to derive its name from its conspicuous scales, it inhabits Haweswater in Cumberland, Ullswater in Westmorland, and the Red Tarn, a small lake on the side of Helvellyn, at a height of 2356 feet; all three lakes belong to the system of the Eden. The Schelly is so closely related to the Powan that many authors have considered them identical.

This form is distinguished by small blackish spots on the back and sides, which are in some specimens few and indistinct, in others very numerous, and present also on the head and in transverse series on the dorsal and caudal fins.   The body is usually deeper and the interorbital region broader than in the Powan; the branched rays in the anal fin number ten to twelve (nine to eleven in the Powan) and the scales round the caudal peduncle twenty-two to twenty-four (twenty to twenty-two in the Powan).

Accounts of the habits and food of this fish are very similar to those which have been given for the Powan.   It is scarce in Ullswater, but still abundant in Haweswater, where it is netted; it reaches a length of 15 inches.

THE GWYNIAD (*Coregonus pennantii*), of Bala Lake in Merionethshire, takes its name from the Gaelic *gwyn*, white; it has a somewhat larger eye than either the Powan or the Schelly; it further differs from them in that there are usually nine or ten (rarely eight) longitudinal series of scales on each side between the lateral line and the base of the pelvic fin, whereas the Powan and Schelly have seven to nine, the latter number being rare.   The anal fin has eleven to thirteen branched rays. The Gwyniad attains a length of 16 inches; its habits appear to resemble those of the Powan.

The Powan, Schelly, and Gwyniad are represented by several very similar forms in the lakes of Scandinavia and Central Europe, including the " Blaufelchen " (*C. wartmanni*) of the Lake of Constance.   A migratory arctic species (*C. muksun*),

PLATE XVII

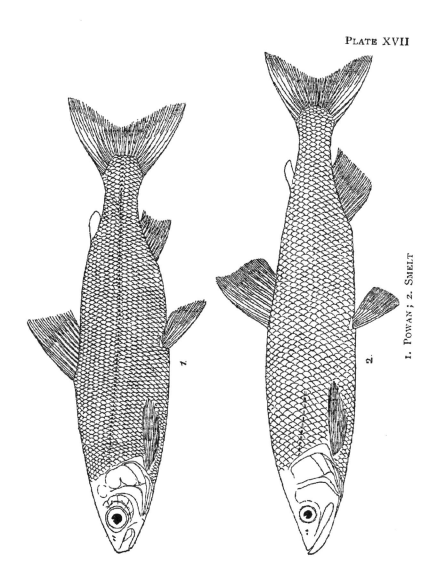

which runs up the rivers of Siberia, is also closely related to these fishes.

THE HOUTING (*Coregonus oxyrhynchus*) is at once recognizable by means of the produced pointed snout; its name appears to be a Dutch corruption of Whiting. It is a marine fish, common from Scandinavia to Holland, and is often seen on the London market. It spawns in fresh water, and occasional examples have been taken in British

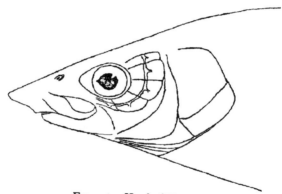

FIG. 15 —Head of Houting.

rivers; three recorded by Day were from Lincolnshire, the Medway, and Chichester, and since then several have been captured in the Colne. The Houting grows to a length of 16 inches.

THE GRAYLING (*Thymallus thymallus*) belongs to a genus especially characterized by the rather long dorsal fin, with seventeen to twenty-five rays, the first four to ten of which are simple. As in the Whitefish the scales are moderately large, whilst in the size of the mouth and the strength of the teeth the fishes of this genus are intermediate between the Trout and the Whitefish.   There are about seven closely related species inhabiting North America, Europe, and northern Asia, especially in arctic or alpine streams.

Our species has the body rather elongate and moderately compressed, with the profile convex from the snout to the dorsal fin; the head is short and subconical, with the mouth subterminal and rather small, the maxillary about reaching the vertical from the anterior edge of the eye; the teeth are small, pointed, in a single series in the jaws, on the anterior part of the vomer and the palatines; the diameter of the eye measures about one-fourth of the length of the head; the dorsal fin is longer than high, with the free edge convex.

The coloration is very variable, usually greenish brown or purplish on the back, silvery grey on the sides, and white below; on the sides a number of dark longitudinal stripes mark the limits between the series of scales, and little scattered blackish spots are usually present also; the dorsal fin is barred with from three to five rows of bluish-black spots.   It is from the greyish colour that the fish takes its name.

The Grayling is abundant in the streams of Lapland, Finland, and other parts of the north of Europe, and is common in Switzerland and in the alpine districts of the surrounding countries, ranging southwards to northern Italy.   In England and Wales the Grayling is locally abundant, occurring in many streams in the northern and midland counties, also in the Dee, the Severn and its tributaries, and the Avon and Itchen in Wiltshire and Hampshire; it has been introduced into Scotland, but is absent from Ireland.

PLATE XXVIII

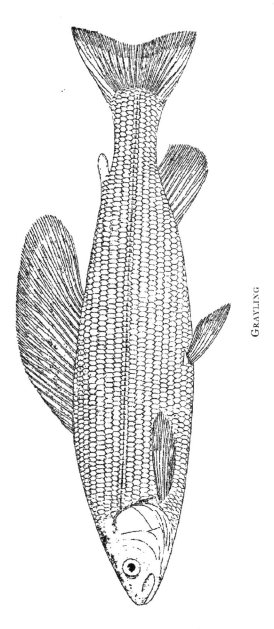

GRAYLING

In our waters this species exceptionally attains a weight of 4 or 5 lbs., but is said to grow to more than twice this size in northern Scandinavia; the example figured (Pl. XVIII) is a foot long.

All our Grayling rivers hold Trout, but not every Trout stream is suitable for Grayling, which prefer clear rapid streams, with alternating pools and shallows and a bottom of clay, gravel, or sand. In the summer-time they may often be seen in considerable numbers lying on the shallows, but the larger fish are said to frequent the deeper water. Their diet includes shrimps, little molluscs, insect larvæ, etc., and they often feed at the surface on flies.

The breeding-season is from March to May, when the fish swim in pairs to gravelly shallows, where they are said to scoop out a hollow for the eggs with the caudal fin, and to cover them over after impregnation in much the same way as the Salmon; the eggs hatch in about a fortnight.

The Grayling is a good sporting fish, of beautiful appearance, and as food even better than Trout, the flesh being white, firm, and of delicate flavour; it is at its best in the autumn, when the Trout are out of season.

CHAPTER VII

THE SMELT AND THE SHADS

The Smelt—characteristic features—distribution—size—food and habits—spawning—rate of growth—sometimes a permanent freshwater resident—as food—meaning of names. The Herring family : the Allis Shad—characters of the species—distribution —size—migrations—spawning—growth of young fish—value as food . the Twaite Shad—differences from the Allis—distribution —size—habits—migrations—spawning—hybrids with the Allis —growth—as food—Shad sometimes permanent freshwater residents

THE SMELT or SPARLING (*Osmerus eperlanus*) has the external features characteristic of the Salmon family, but, from certain peculiarities of the cranial structure, the dentition, and the soft anatomy, it may be placed in a distinct family (*Argentinidæ*), which includes several genera and species from northern seas.

The Smelt has the body fusiform, compressed, and elongate, the head pointed, with the eyes of moderate size, and the interorbital region nearly flat. The mouth is wide and oblique, with the lower jaw curved upwards in front and projecting; the teeth in the upper jaw are feeble, but in the lower they are strong canines; similar strong teeth are placed on the tongue, the vomer, and the palatine bones , the most characteristic feature of the dentition, however, is the presence of a series of teeth on each side of a median longitudinal groove in the roof of the mouth, these are situated along the inner edges of the mesopterygoid bones, and are found not only in the other members of the family, except some degenerate deep-sea forms, but also in the allied fishes (*Retropinnatidæ, Haplochitonidæ, Galaxiidæ*), which take the place of the Salmonids and Smelts on the

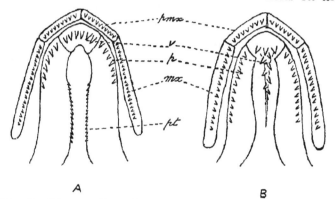

A   B

FIG. 16.—Diagrams illustrating the upper dentition of a Smelt (A) and a Trout (B).

pmx. præmaxillary; *mx.* maxillary; *v.* vomer; *p.* palatine; *pt.* mesopterygoid.

coasts and in the rivers of Australia, New Zealand, and Patagonia. The scales are thin, transparent, and rather large, numbering sixty to sixty-five in a longitudinal series; the lateral line runs on about ten scales only, and then ends; the dorsal fin, of two or three simple and eight or nine branched rays, originates above the pelvics, whilst the anal is rather long, the branched rays numbering twelve to fourteen. The coloration is silvery, with the back olive-green.

There are about six species of Smelts; ours is

principally an inhabitant of the Baltic and the North Sea, ranging southwards on our coasts as far as Hampshire and in France to the Seine; on the western coasts of Britain it is found southward to North Wales, and it probably occurs also on the northern coasts of Ireland. It attains a length of 13 inches and a weight of ½ lb.; a specimen nearly this size is figured (Pl. XVII, Fig. 2).

This is a greedy fish of prey, feeding on small fishes, shrimps, worms, etc.; in the early spring the Smelts assemble in shoals and ascend the rivers to spawn, in some localities not going farther than tidal waters for this purpose, but in others pushing up well beyond. The spawning takes place usually from March to May, when the fish crowd together in dense array and the eggs are shed; the latter attach themselves wherever they happen to fall; thus in the Forth, where the Smelts spawn in March, about two miles below Stirling Bridge, every stone, plank, and post has been described as covered with their yellow eggs. After spawning the Smelts usually return to the sea, but their habits are very variable, as in some rivers they stay for months, in others not more than two or three weeks.

The eggs hatch out in from one to three weeks, according to the temperature; the fry attain a length of from 2 to 3 inches before the autumn, when they descend to the sea, returning to the estuaries in the spring; in their second season they attain a length of from 4 to 6 inches, and may spawn for the first time in the following spring.

In many Swedish lakes the Smelt is a freshwater resident throughout the year, and in Britain this seems to be the case in Rostherne Mere in Cheshire, whilst they have been found to thrive and breed when kept in ponds.

The Smelt is a valued food-fish, and is in best condition for the table in the autumn and early spring. It has been wrongly said to take its name from its odour, resembling that of cucumbers, and sometimes quite strong; but there can be little doubt that the word Smelt is from the Anglo-Saxon *smeolt*, smooth and shining. Sparling is the equivalent of the German *Spierling* and the French *éperlan* (old French *esperlan*).

The Herring family (*Clupeidæ*) is closely related to the Salmonids, differing especially in that both lateral line and adipose fin are absent, whilst two supramaxillaries, instead of one only, are attached to the maxillary bone, and oviducts are present. The fishes of this family occur in large numbers in all tropical and temperate seas, but rarely at any great distance from the coast; many of them enter fresh water, and a few are permanently fluviatile or lacustrine. Our British species are the Anchovy, Herring, Pilchard, Sprat, Allis Shad, and Twaite Shad; the two last spawn in fresh water and must, therefore, be noticed here

The Shads are very similar in most respects to the Herring, but are sometimes placed in a distinct genus (*Alosa*) characterized by the presence of a notch at the extremity of the upper jaw; our two species have been so often confused with each other and with allied species that some details of their characteristic features must be given.

THE ALLIS SHAD (*Clupea alosa*) has the body moderately elongate and strongly compressed; the greatest depth, usually about one-fourth of the length to the base of the caudal fin, but in large specimens sometimes nearly one-third, is not far behind the head; the snout is blunt and rather short, the eye of moderate size, protected by a transparent lid with a vertical elliptical aperture above the pupil; the mouth is terminal and oblique, with the teeth minute or absent and the maxillary extending a little beyond the eye in the adult fish. The gill-openings are wide and the gill-rakers are

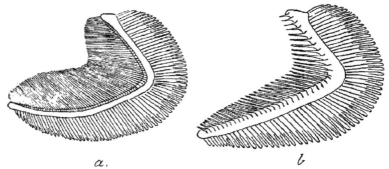

a.   *b*

FIG. 17.—Anterior branchial arches of Allis Shad (*a*) and Twaite Shad (*b*), showing the difference in number and length of the gill-rakers.

slender, very long, and in the adult fish very numerous; in specimens from 1 to 2 feet long I count seventy-two to eighty on the lower limb of the anterior branchial arch, and in one of 8 inches sixty; in smaller examples the number is still less. The scales are irregular and deciduous and, therefore, not very easy to enumerate; I count seventy-two to eighty-six in a longitudinal and twenty-one to twenty-five in a transverse series; the trenchant ventral edge of the body is serrated, bearing a series of bony scutes with backwardly directed points.

The dorsal fin, placed about in the middle of the length of the fish, is composed of four or five graduated simple and fourteen to eighteen branched rays; the anal has three simple and twenty to twenty-four branched rays, and its base usually measures about one-fourth of the distance from the head to the caudal fin; the latter is deeply forked; the pectorals almost reach the pelvics in the young, but not in the adult, whilst the small pelvics are inserted nearly below the origin of the dorsal. The coloration is silvery, with the back greenish or purplish; there is usually a dark spot on the shoulder, which is sometimes followed by a series of smaller spots.

The Allis Shad is found on the Atlantic coasts of Europe and in the western Mediterranean; it attains a length of about 30 inches, and a weight of 8 lbs.

In the sea this fish swims probably in small companies, feeding especially on little crustaceans, but at times also on the fry of other fishes. In the spring, usually about April, the shoals ascend the rivers, sometimes for enormous distances, thus in the Rhine they reach Switzerland, and in the Elbe Bohemia; they also penetrate nearly to the sources of the Seine and Loire. They do not seem to be very plentiful in our rivers except the Severn and the Shannon; heavy floods or thunderstorms check their ascent, and may even drive them back to the sea; once well clear of the estuaries and fairly started on their journey they appear to cease feeding and to devote all their energies to pressing on towards their spawning places; they usually breed in May or a little later (hence the German name *Maifisch*), and make

10

a considerable commotion, swimming rapidly at the surface, and pressing together and thrashing the water with their tails during the process of deposition and impregnation of the eggs, which sink and lie free at the bottom. The fish are so exhausted after breeding that many perish; others get back to the sea, so that by the end of the summer none are to be seen.

The eggs are small and very numerous; the young fry feed on minute crustaceans, and reach a length of 3 or 4 inches before the winter, when they disappear, doubtless hibernating at the bottom, and perhaps hiding under stones; during the next spring and summer they attain a length of 6 or 7 inches in fresh water, and about July go down to the sea, whence they next reappear as adult fish, rarely less than a foot long and probably three or four years old. The young fry found in fresh water have fewer gill-rakers than the adult fish, usually from thirty to forty on the lower part of the anterior branchial arch.

The Allis Shad is in best condition for the table from the time when it first enters fresh water until it is ready to spawn; the flesh is said to be excellent, although the numerous bones may be troublesome.

THE TWAITE SHAD (*Clupea finta*) has the body on the average a little more slender than the Allis Shad, the greatest depth being contained three and a half to four and one-fourth times in the length of the fish; also the shape is more regularly fusiform, the highest point of the body being at the origin of the dorsal fin. The gill-rakers are much shorter and fewer,

Plate XIX

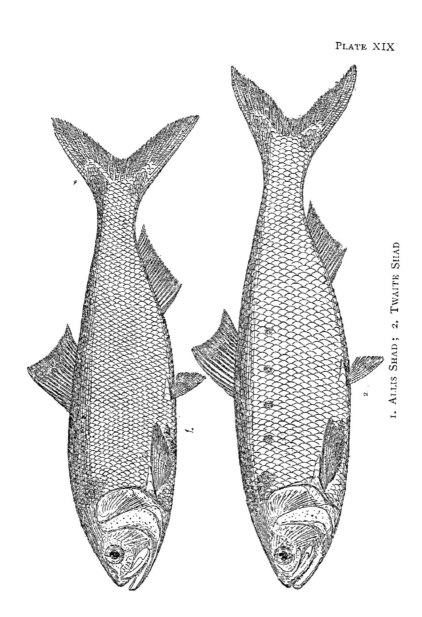

1. Allis Shad; 2. Twaite Shad

numbering twenty-four to thirty on the lower part of the anterior branchial arch in fish of more than 6 inches long. The scales are larger and more regularly arranged; I count fifty-eight to sixty-six in a longitudinal and sixteen to twenty in a transverse series. The specimens I have examined have twelve to sixteen branched rays in the dorsal and sixteen to nineteen in the anal fin, whilst the base of the latter measures about one-fifth of the distance from the head to the caudal fin; the pectorals seem to be somewhat shorter than in the Allis Shad. The external differences between this species and the preceding are shown on Plate XIX, where two fish of nearly the same length, 11 and 12 inches, are figured. This species occurs on the Atlantic coasts of Europe, and is represented in the Mediterranean by an allied but distinct form (*Clupea nilotica*); it attains a length of 20 inches and a weight of $3\frac{1}{2}$ to 4 lbs.

In its habits the Twaite closely resembles the Allis Shad, but is generally smaller and occurs in greater numbers; from the structure of its gill-rakers one would infer that the minute crustaceans, which form so large a proportion of the diet of its congener, are not consumed by it to the same extent, and that it is perhaps more destructive than the Allis Shad to the fry of Herrings, Sand-eels, etc.

This species usually enters fresh water later than the Allis Shad, from the middle of April to the beginning of June; in the Rhine it does not ascend nearly so far as its congener, but in our rivers, which are much smaller, there is little to choose between them in this respect. Thus the Twaite used to ascend the Severn as far as Montgomeryshire, although it is now unable to get beyond Worcester. In May it appears in the lakes of Killarney, and I have little doubt that the following passage from a History of the County of Westmeath, written in 1682 by Sir Henry Piers, refers to this fish: "About Lough Tron and Lough Direvragh there is found, in the month of May only, a small fish, without spot, of almost the same shape as a herring—a fish very pleasant and delightful, but not taken in great quantities; the natives call it Goaske."

Although the time of spawning of this species is usually a little later than that of the Allis Shad, there is a considerable overlapping, as is shown by the not infrequent occurrence of hybrids, easily recognized by the intermediate number of gill-rakers.

The growth of the Twaite in fresh water is precisely similar to that of the Allis; the fry of 3 to 6 inches have fewer gill-rakers than the adult fish, seventeen to twenty-four on the lower part of the anterior branchial arch. As food the Twaite is held in somewhat less estimation than the Allis Shad.

It may be of interest to mention that Shad, like Salmon, Trout, and Smelts, can live and breed in fresh water without going to the sea; the large lakes of northern Italy are inhabited by a species of Shad (*Clupea lacustris*) which is a permanent resident. This is a small and slender form, with the number of scales and fin-rays as in the Twaite, but with the gill-rakers not much less numerous than in small Allis Shad; in three from Lake Garda, 4, 5, and 6 inches long, I count respectively twenty-eight, thirty-two, and thirty-six gill-rakers on the lower limb of the anterior branchial arch.

Students of the biology of our two species of Shad should consult an interesting memoir by P. P. C. Hoek, entitled " Neure Lachs—und Maifisch —Studien," published in *Tijdschr. Nederl. Dierk. Vereen.* (2), vi. Leiden, 1900.

CHAPTER VIII

THE PIKE

Relatives of the Pike. Distribution of our species : specific characters—size—food and habits—greediness sometimes causes death—Pike rushing into the jaws of another in pursuit of prey—cannibalism—attacks on large animals and man—supposed dislike of Tench—pairing—breeding—habits of fry—rate of growth—age—as food—names and their derivation

THE PIKE (*Esox lucius*) is a member of a small order (*Haplomi*), well defined by osteological characters, which contains in addition to the Pike family the Mudfishes (*Umbridæ*), small fishes of stagnant pools, swamps, and ditches, one species inhabiting Austria - Hungary, the other North America, and the Blackfishes (*Dalliidæ*), comprising a single species in Alaska and Siberia, a curious fish of extraordinary vitality, remaining frozen for weeks in the winter and thawing out as lively as ever.

The Pike family includes only one genus (*Esox*), with five species, all found in North America. In the United States the three southern species are the smallest, and are known as Pickerel, whilst the giant Maskinongy of Canada grows to a length of 8 feet and a weight of more than 100 lbs.

Only one species occurs in the Old World at

PLATE XX

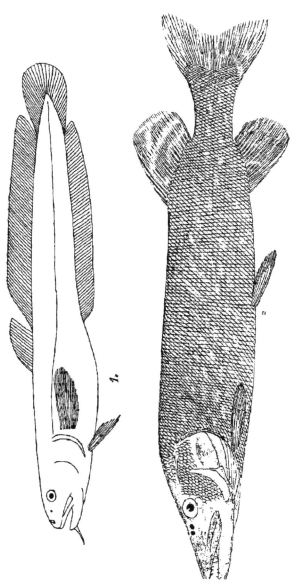

1. BURBOT; 2. PIKE.

the present day, although two or three have been found fossil in Oligocene and Miocene deposits in Central Europe. Our species is found all over Europe, except Spain and Portugal, and in Russian Turkestan, Siberia, and Mongolia ; in North America it ranges from the region of the Great Lakes to Alaska. It is abundant in all four countries of the British Isles, but becomes rather scarce and local in the Northern Highlands of Scotland.

The Pike resembles the members of the Herring and Salmon families in that the maxillaries border the mouth, all the fin-rays are flexible and jointed, the pectoral fins are placed low down, and the pelvic fins are abdominal in position ; important differences are the separation of the præmaxillaries by the vomer and the backward position of the dorsal fin, which is opposite to the anal.

The body is elongate, covered with small scales, the head flat above, the snout prolonged and depressed, and the mouth large; the eyes are placed high, about in the middle of the length of the head. The præmaxillary teeth are small and the maxillaries are toothless; the roof of the mouth is furnished with three parallel bands of slender, pointed, backwardly directed, depressible teeth, the middle band on the vomer, the lateral ones on the palatine bones ; there is also a band of teeth on the tongue; the strongest teeth are at the sides of the lower jaw, where they are few, fixed and erect, as eminently adapted for seizing prey as the rest are for facilitating its passage inwards and preventing its escape.

In colour the Pike is greenish, with the lower parts white, and with yellow spots or undulating bands on the sides; the dorsal, anal, and caudal fins are olive-green, with blackish spots or stripes.

The size attained by the Pike in our islands has been the subject of a good deal of controversy, and according to one authority "more lies have been told about the Pike than about any other fish in the world." Well-authenticated instances of the capture of Pike of from 35 to 45 lbs. weight are plentiful, and there are many tales of much larger fish, which may be true enough, but unfortunately cannot be verified. Ireland has always been celebrated for the size of its Pike, and a story is told with plenty of circumstantial detail of the capture of one of more than 90 lbs. weight in the Shannon about a century ago. Thompson says: "The Rev. C. Mayne, writing from Killaloe in 1838, gave me the names of two gentlemen who killed Pikes of 49 and 51 lbs. weight in that locality, and also informed me that in August, 1830, Mr. O'Flanagan (then aged seventy) killed with a single rod and bait, in a lake in the County Clare, a Pike of 78 lbs."

It is generally considered that the best-authenticated record of a monster is that of the famous Kenmure Pike, taken towards the end of the eighteenth century in Loch Ken by John Murray, gamekeeper to Viscount Kenmure. As the captor bore it along to his master, with the head over his shoulder, the tail swept the ground. According to the Rev. W. B. Daniel (*Rural Sports*, 1801–1813) this fish weighed 72 lbs., and was caught with a fly made of a peacock's feather. Another author, Dr Grierson, published some "Mineralogical Observations in Galloway," in the *Annals of Philosophy for*

1814, and in a foot-note wrote. " I have very often killed in Loch Ken Perch weighing 4 lbs., and at one time a Pike of 7 ; but this is nothing in comparison of one that was caught about forty years ago in this lake by John Murray, gamekeeper to the Hon. John Gordon of Kenmore. It weighed 61 lbs., and the head of it is still preserved in Mr. Gordon's library at Kenmore Castle."

Although these early writers were not in agreement as to the actual weight of the fish, both stated that the head had been kept, and indeed the greater part of the skeleton of the head is in Kenmure Castle at the present day, to bear witness to the fact that a Pike of exceptional size was once captured. The head was evidently severed from the body too far forward, so that the occipital region of the skull is wanting and the opercular bones are incomplete, but measurements I have made indicate that the fish probably weighed as much as 61 lbs., if it was in good condition, and possibly even 72 lbs.

These measurements are identical with those of the head of another Loch Ken Pike in the possession of Sir Arthur Henniker-Hughan, Bart., in his house at Parton. This fish was taken in the summer of 1904, when a porter at Parton Station saw it stranded near the edge of the lake and lifted it out of the water; it was in an emaciated condition, and weighed only 39 lbs.

The enormous difference in weight between a ripe and a spent fish is well known, and is strikingly exemplified by two casts of Pike in the Buckland Collection, very similar in all their proportions except depth and girth; the one captured in March

weighed 32 lbs., the other taken in May only
20½ lbs.   Still greater is the difference between a
fish in good condition and an emaciated fish which
has perished of starvation or from senile decay,
and in which the head appears disproportionately
large.   So that the weight of Sir Arthur Henniker-
Hughan's fish may be taken as corroborative
evidence that the Kenmure Pike actually weighed
some 60 or 70 lbs.

The principal measurements in inches of the
heads of three large Pike from Loch Ken may be
of interest.   A and B are the property of Sir
A. Henniker-Hughan ; A pertains to a Pike caught
in June, 1898, which weighed 37 lbs., and would
doubtless have weighed considerably more a few
months earlier or later.   B is the head of the
39-lb. Pike mentioned above, and C that of the
Kenmure Pike.

	A	B	C
Total length (measured in a straight line from level of end of snout to extremity of operculum)	11	12	—
Length on upper surface from anterior edge of vomer to posterior edge of frontals	7½	8½	8½
Greatest width across frontal bones	2¾	3	3
Extreme length of maxillary	4⅝	5	5
Lower jaw (measured in a straight line from symphysis to angle)	8¼	8¼	8¾

These measurements are either of the skeleton of
the head (C) or of heads which have been dried
(A and B), and are less than they would have been
if the fish had been fresh or preserved in spirit.
The figures scarcely convey an adequate idea of
the great difference in bulk between the heads A
and B ; no one who compared them side by side

could doubt that the latter pertained to a fish far
heavier, under normal conditions, than the owner
of the former, especially if he bore in mind the
fact that, other things being equal, the larger the
fish the smaller the head proportionately ; growth
of the head ceases or becomes almost imperceptible
before that of the fish as a whole.

It was recorded in the *Field* that in May, 1905,
a Pike of 48 lbs. was gaffed in one of the inlets of
Lough Corrib, a nearly spent female fish, which it
was thought might have weighed not much less than
60 lbs. had it been captured before it had spawned ;
this estimate was probably nearly accurate, for
Buckland found that the ovaries of a ripe 32-lb.
Pike weighed 5 lbs.

Some measurements of the head of this fish were
given, including the following : length of head,
measured from extremity of lower jaw to end of
operculum, 13 inches ; interocular width, $2\frac{1}{2}$ inches.
These figures are not directly comparable with those
given above, but after examining the heads of
several large Pike I conclude that the head of this
fish was almost as large as those of the Loch Ken
giants, the reputed 72 pounder and the emaciated
39 pounder.

I have seen a 39-lb. Pike from Lough Derg, 48
inches in total length (to the end of the middle rays
of the caudal fin) and with the head about 11 inches
long (about 12 inches if measured round the curve
of the head to the end of the lower jaw). Our
figure (Pl. XX, Fig. 2) is of a " Jack " of about
3 lbs.

Pike are found in lakes and rivers, sometimes
even in quite small ponds or in little streams if

there are pools where they can find shelter; but they only grow large where they have plenty of room and food in abundance.

The Pike leads a solitary life and often seems to select a pool or a stretch of quiet water which he regards as his domain for the season, and from which he never wanders far. On warm days in the summer he may often be seen lying motionless, basking at the top of the water and resembling a log of wood, or he may lurk within the shelter of a clump of lilies or a bed of reeds, from which concealment he may suddenly dash out to seize his prey. Often the Pike is not hungry, and takes no heed of the little fish which sport round him, unconscious of their peril; but he makes up for these "off days" when he is really on the feed, which in my experience is especially when a good westerly breeze is blowing. At such times he goes in pursuit of the shoals of Roach, Dace, Gudgeon, etc., and everything that comes in his way is welcome, whether it be a member of his own or another species of fish, a frog, a water-rat, a moor-hen, or a duck. Some writers have computed that a Pike will consume in one day his own weight of food, so insatiable is his appetite and so rapid his digestion.

Fish are, of course, his normal diet, and these he seizes crosswise and then swallows them head first. I well remember fishing with my father on the Stour, near Stalbridge, and hearing a considerable splashing a little way off; we ran to the spot and saw an enormous Pike, carrying a fish crosswise in his mouth, swimming round and round in a circle at the top of the water quite close to the bank on which we stood; he continued doing this for perhaps a minute before sinking out of sight.

Innumerable instances might be adduced to illustrate the fierceness and greediness of the Pike Smitt relates a story of a Pike of 7 or 8 lbs. which caught and ate a Salmon of its own size, darting forward and seizing its victim right across the body; a fierce combat ensued, but the Salmon could not shake off its relentless captor, and in a couple of hours was so exhausted that the Pike began to swallow it head first. The meal lasted three days before the whole body had disappeared, and for a week afterwards the Pike had a very swollen appearance, and could hardly be induced to move by touching it with a long stick.

This Pike was perhaps more fortunate than he deserved to be, for it has often happened that the attempt to swallow too big a prey has so exhausted or choked the captor as to cause his death. Some years ago in one season two Pike, weighing 35 and 29 lbs. respectively, were found floating dead on the lake at Sherborne in Dorset; each had failed to swallow a fish of about one-third its own weight, a Carp in the one case, a Pike in the other. On this same lake I have seen a Pike rush with such fury at a Roach that was just being lifted out by an angler that only the head was left on the hook.

On Loch Tay in 1870 two Pike, of nearly equal size, were seen struggling with the head of one in the mouth of the other; they were gaffed by a boatman and were found to weigh together 19 lbs.; these two fish are preserved as a cast in the Buckland Collection. This incident is perhaps explained by the following, narrated by Mr. Pennell: " A ludicrous circumstance once happened in the feeding of two Pike kept in a glass vivarium   A bait was thrown in about midway between the fish, when each simultaneously darted forward to secure it, the result being that the smaller fish fairly rushed into the open jaws of the larger, where it remained fixed, and only extricated itself with difficulty and after a lapse of some seconds."

Cases of undoubted cannibalism are numerous enough, and a typical one is mentioned by Mr. Pennell in his *Book of the Pike*.   A gentleman had set a trimmer overnight in the River Avon at Chippenham, and on proceeding the next morning to take it up he found a heavy Pike apparently fast upon his hooks; in order to extract these he opened the fish and then saw inside it another Pike of considerable size, from the mouth of which the line proceeded.   On opening the latter fish a third Pike of about three-quarters of a pound weight, and already partly digested, was discovered in its stomach; this last fish was, of course, the original taker of the bait.

Even large animals are sometimes attacked by the Pike, and there are plenty of modern instances to parallel those contained in the following passage from Walton: " Gesner relates a man going to a pond, where it seems a pike had devoured all the fish, to water his mule, had a pike bit his mule by the lips; to which the pike hung so fast, that the mule drew him out of the water, and by that accident the owner of the mule angled out the pike. And the same Gesner observes, that a maid in Poland had a pike bit her by the foot, as she was washing clothes in a pond.   And I have heard the like of a woman in Killingworth pond, not far from Coventry. But I have been assured by my friend Mr. Seagrave, of whom I spake to you formerly, that keeps tame otters, that he hath known a pike in extreme hunger fight with one of his otters for a carp that the otter had caught, and was then bringing out of the water."

Mr. Pennell gives a well-authenticated account of an attack made by a Pike on a boy of fifteen, who was bathing with his companions in a pond near Ascot in June, 1856. Both hands were seized and severely bitten; a few days afterwards the fish was seen floating on the water almost dead; it was taken out and was found to measure 41 inches in length, but was very lean, and was evidently dying of starvation; we may conclude that before coming to this pass he had eaten all the other fish in the pond.

Sufficient has been said to indicate the voracity and ferocity of the Pike, which when driven by the pangs of hunger will even attack man himself; anger and revenge also seem to influence him, for I have seen one spring from the ground and close his jaws on the arm of the angler who was about to take the hook from his mouth. It has also been credited with one good trait (in addition, as Day remarks, to eating its younger relatives), and that is abstention from devouring the Tench out of gratitude for services performed by the latter; the Tench is popularly regarded as the physician of the Pike, whose wounds it is said to heal by means of the slime which covers its body. There is some evidence that Pike, as a rule, do not care for Tench, although in some localities they are known to take them; but that

11

this dislike is inspired by gratitude is, to say the least of it, highly improbable.

Towards the winter the Pike forsake the weedy shallows for deeper water and begin to pair, choosing for mate a fish of nearly the same size, or if either be the larger it is usually the female; in January I have seen a male and female fish of exactly the same size ($5\frac{1}{2}$ lbs.) caught one after the other in the same place, and Mr. Pennell says that he has even met with a happy couple quite early in the autumn. In March or April, or sometimes as early as February, they leave the open water and make their way into the ditches and backwaters or on to any quiet shallow, where they spawn among the weeds. The smaller fish spawn first, the larger ones not being ready until later; sometimes two or three males have been observed in attendance on one female. After breeding they return to the lakes and rivers, but take some time to recover, so that the larger fish are not really in good condition before the autumn.

The eggs are numerous (Buckland estimated that there were 595,200 in a Pike of 32 lbs.) and usually hatch out in from ten days to three weeks, in about another fortnight or less the yolk-sac is absorbed, and the fry begin to feed, at first on larvæ, insects, shrimps, etc., and soon afterwards on the fry of other fish, growing very rapidly. Many are said to die through attempting to swallow Sticklebacks, which erect their spines and stick in the Pike's throat.

In the summer I have often walked along the side of the little river at Sherborne, in a place where the current of the stream is separated from the bank by a bed of weeds, which almost reach the surface of the water; at every few steps one would come across a little break in the weeds, and in each one a little Pike, perhaps 4 or 5 inches long, and three or four months old, which, when disturbed, would shoot out towards the main stream as far as the edge of the weeds. As Lubbock has observed in Norfolk, the Pike is solitary from the first, and in summer almost every distinct puddle in fens where turf is cut has its tiny tyrant in an infant Pike, who enacts despotic sovereignty and lords it over tadpoles and fry.

During a good many seasons' fishing in Dorsetshire on the Stour in August and September I have caught a number of Pike of about $\frac{3}{4}$ lb., $1\frac{1}{2}$ lbs., and $2\frac{1}{4}$ to 3 lbs. weight, and there can be little doubt that these were rather more than one, two, and three years old respectively; this seems to be about the usual rate of growth in most good streams, and afterwards for some years the increase is said to be about 2 lbs. a year, but there is some very good evidence that in favourable circumstances growth may take place at the rate of as much as 4 or 5 lbs. a year. However, after a certain size has been reached the rate of growth must decrease, and if the larger the Pike the later the spawning and the longer the period required to get into condition, it follows that a time must come when growth will cease, as the fish will be able to do no more than make good his losses, and this will be succeeded by a period of senile decay, when the Pike, aged and worn out, becomes lanky and large headed, and if in his enfeebled condition he escape his cannibalistic relatives may ultimately die simply from inability to capture food or to assimilate it. Such was probably the history of the 39-lb. Loch Ken Pike described above.

Nothing definite can be said as to the age to which a Pike may live, although it is probable that fish of 60 or 70 lbs. weight are at least as many years old. But although Pike are reputed to live to a very great age there are not, so far as I am aware, any satisfactory proofs that they actually do so.

A remarkable fact in the life-history of the Pike is the way in which it spreads, and Mr. Pennell inclines to the hypothesis that in wet weather it may travel overland from one pond to another, and gives some curious instances in support of this. Probably the appearance of Pike in unexpected places is often due to the fact that the young are hatched out in ditches at a time when these are full and temporarily connected with pieces of water which are afterwards isolated; a rut full of water is almost enough for the passage of an infant Pike.

Nowadays Pike are generally not held in much repute as food, but a river fish of medium size taken in the autumn or winter is by no means bad if properly cooked. In the Middle Ages they were esteemed as food-fishes, and were a feature of every banquet; Macpherson tells us that in 1298 Edward I authorized Robert de Clifford to let the Bishop of Carlisle have sixty jacks to stock the moats of Carlisle Castle, whilst Walton, after giving elaborate directions how to prepare and cook a Pike, quaintly says, "This dish of meat is too good for any but anglers, or very honest men."

The name Pike seems to be derived from the Saxon *Piik*, pointed, in reference to the form of the head, whilst the alternative name Luce is from the Latin *Lucius*. The Lowland Scotch Gedd is the equivalent of the Scandinavian *Gadda*, of much the same meaning as Pike, 'Jack' is commonly applied to small Pike of not more than 3 lbs. weight.

CHAPTER IX

THE EEL

The order *Apodes*. The Eel family : description of the Eel—
its distribution—differences from the American Eel—size
and weight of Eels—only one European species—Yellow Eels
—Frog-mouthed or Broad-nosed Eels—food and habits—
journeys across country—Silver Eels—differences from Yellow
Eels—migrations to the sea—ripe Eels—spawning—Lepto-
cephalids—the larva of the Eel—metamorphosis and migration
—Elvers—age and rate of growth—distribution of Eels
dependent on proximity to suitable breeding-places—bibliography of biology of the Eels—value as food

THE order *Apodes* includes fishes which agree
with those Teleosteans already considered
in the persistence of the duct of the air-bladder,
the absence of spinous fin-rays, and the abdominal
position of the pelvic fins, when present ; however,
in all the living members of the group these fins
are absent, and they are also characterized by their
more or less snake-like form and small gill-openings,
but especially by the structure of the mouth, which
is bordered above by the maxillaries laterally and
in front by the vomer, the præmaxillaries being
absent, or at any rate not present as separate
elements.

Nearly all the fishes of this order are marine,
and the Eel family (*Anguillidæ*) includes a number

of species, most of which are inhabitants of the deep sea, the genus of the Common Eel (*Anguilla*) differing from the majority in that its members pass the greater part of their life in fresh water or in shallow water near the shore, and only revert to their original habitat in the depths of the sea for the purpose of breeding.

The Eel (*Anguilla anguilla* or *A. vulgaris*) has the body long and subcylindrical, covered with small oblong scales embedded in the slimy skin and arranged in little groups, which are placed obliquely and at right angles to each other; the mouth is terminal, furnished with bands of pointed teeth in the jaws and on the vomer; the small gill-openings are placed just in front of the pectoral fins; the dorsal and anal fins are long and confluent with the caudal; the first - named commences at some distance behind the head, but well in advance of the anal.

The Eel is found in the Mediterranean and the rivers which drain into it, and on the Atlantic coasts and in the rivers of the Atlantic slope of Europe as far north as Scandinavia, but it is absent from the Black Sea and its tributaries; a very similar species (*A. chrysypa*) is found on the American side of the North Atlantic, which differs from our Eel chiefly in the fewer vertebræ (104 to 110 instead of 111 to 118) and in the more posterior origin of the dorsal fin, the distance between the beginnings of the dorsal and anal fins being always less than the length of the head, measured to the gill-opening, in the American Eel, whereas in our species it is about equal to the length of the head.

The female Eels grow to a much larger size than the males; the latter do not often attain a length of much more than 20 inches, whereas female examples 3 feet long and weighing 4 or 5 lbs. are by no means uncommon, whilst a length of 5 feet and a weight of 12 to 15 lbs. is sometimes attained. In the Buckland Collection there is the cast of an Eel from the Mole, which measured a little more than 4 feet 6 inches in length and weighed 10 lbs., and Walton may be believed in his statement that an Eel 1¾ yards long was caught in the river at Peterborough in 1667. The Eastern counties seem to be the home of the largest Eels, for the Rev. R. Lubbock's record of one of more than 20 lbs., taken near Norwich in 1839, can scarcely be questioned, whilst Yarrell saw at Cambridge the skins of two said to have weighed 23 and 27 lbs., which were taken on draining a fen-dyke at Wisbech.

The Eel is subject to great variation, and on this account has received a number of names, whilst expert zoologists like Yarrell and Thompson recognized three or four species in the British Isles alone. It was not until 1896 that Petersen, a Dane, showed the reason of this variation and established beyond doubt that there is only one species of European Eel.

Most Eels may be referred to one of two types, i.e. Yellow Eels, or Eels in their growing dress, and Silver Eels, or Eels in their breeding dress (cf. Pl. XXI). The Yellow Eels are usually rather pale, with the back greyish, brownish, or greenish, the sides yellow, and the belly similarly coloured, or white; the eyes are small and more or less turned

PLATE XXI

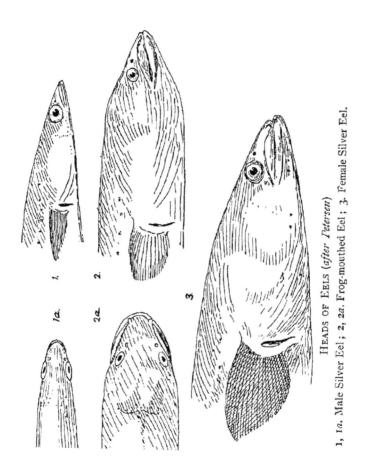

HEADS OF EELS (*after Petersen*)

1, 1*a*. Male Silver Eel; 2, 2*a*. Frog-mouthed Eel; 3. Female Silver Eel.

upwards, the snout is flat, and the pectoral fins are pale and rounded behind. The males cannot readily be distinguished from the females without dissection, and the sexual organs are little developed. Many of the larger Yellow Eels become very voracious and, principally in consequence of the excessive development of the jaw-muscles, acquire a very characteristic appearance; from this they have received a number of names such as " Frog-mouthed Eel," " Bulldog-headed Eel," etc., whilst other names such as " Glut-eel," " Gorb-eel," and " Hunter Eel" refer to their predaceous and gluttonous habits.

According to Petersen, such " Frog-mouthed Eels " are abundant in both fresh and salt water, and are nearly always females, they are often captured on hooks baited with small fish. Thompson mentions this form under the name " Broad-nosed Eel " as occurring in Lough Neagh, where it feeds chiefly on the Pollan and destroys them in the nets, and is taken in the summer with night-lines baited with large worms or small Perch; the fishermen called it *Collach* (wicked), and considered it coarse to eat compared with other Eels; some had the head nearly semicircular in outline, and were called Bulldog-headed Eels.

So different is this Frog-mouthed Eel from the normal Yellow Eel, which is sometimes called a " Snig," that even Dr. Gunther considered it to be a distinct species (*Anguilla latirostris*); his example from the Itchen is 2 feet long, and is noticeable for the large head, longer than the distance between the origins of the dorsal and anal fins, the large mouth, the cleft of which extends to the vertical from the posterior edge of the eye, the thick lips, the obtusely conical teeth forming broad bands, the small eyes, their diameter measuring two-fifths the length of the snout, or one-half the interocular width, and the short and blunt pectoral fins.

Yellow Eels, then, are Eels which are feeding and growing, and may vary in length from a few inches to more than 5 feet; they are found both in fresh and salt water, in the former case inhabiting not only rivers and lakes, but small brooks and isolated ponds, in the latter dwelling in harbours or near the mouths of rivers, or lurking among rocks and weeds along the shore. In winter they usually withdraw into deeper water and lie torpid, buried in the mud, but are susceptible to very severe weather; thus Thompson records that in 1841 a large number were killed in the month of February by protracted hard frosts with strong easterly winds, and floated down the Lagan to Belfast.

At other seasons they feed principally at night, in the daytime lying still in holes or beneath stones, or buried in the sand with the head just projecting; during thunder they become restless and active. They feed chiefly on worms, small fish, cray-fishes, etc., but the larger ones are practically omnivorous, and include frogs, water-fowl, water-voles, etc., in their dietary. Not many years ago a large Eel was captured in a pond near Sherborne by a labourer, who noticed a Swan in difficulties and went to see what was the matter; the bird had put its head under water and this had been seized by the Eel, who would not let go until it was in the grasp of the man, who landed it.

It has been asserted that Eels leave the water to feed on slugs, pea-pods, etc., but while this is very doubtful there can be no question that they migrate across country from one piece of water to another on a damp night, being able to live a long time out of water; two small Eels kept by the writer in an aquarium passed most of the day buried in the sand at the bottom, but night after night they made their escape and were always found in the morning on the other side of the room apparently dead; however, when returned to the water they swam about, none the worse for their excursion.

Towards the autumn a certain number of the Yellow Eels, but never any less than a foot long, put on their breeding dress and prepare to migrate to the sea; in other words, they change into Silver Eels. These, which are often termed "Sharp-nosed Eels," differ from the Yellow Eels in that the eyes are larger and not turned upwards, the snout is less flattened, the body is more rounded, plumper, and firmer to the touch, the pectoral fins are longer and more pointed and blackish in colour; the back is blackish and is separated from the silvery white of the lower parts by a lateral stripe of bronze. Internally they differ from the Yellow Eels in that the sexual organs are riper and the digestive tract is shrunken.

The Silver Eels feed but little, and in the late summer and autumn migrate to the sea, and then make their way into the depths of the North Atlantic or of the Mediterranean, if they come from the countries bordering that sea, where they breed. With the practical cessation of feeding the jaw-muscles decrease in size, so that the head of the sometime "Frog-mouthed Eels" loses its abnormal appearance. The adoption of a silvery livery before going down to the sea recalls the parallel case of the Salmon and Sea-trout; the increased size of the eye is no doubt preparatory to life in the deep sea, whilst the larger size and altered shape of the pectoral fins may be connected with the long journey to be taken.

During the migration the Silver Eels are easily captured by Eel-traps, nets, etc., in the rivers and estuaries; according to Thompson, the fishing on the Bann, which flows out of Lough Neagh, used to begin in August, and as many as 70,000 Eels were taken in one night in sugar-loaf-shaped nets placed between the weirs. A run of Eels takes place only on a dark night and can be stopped by bonfires or torches.

Our knowledge of the life of the Silver Eels in the sea is somewhat conjectural; they have been observed on one occasion swimming in numbers near the surface, and have been taken from the stomachs of Whales, Swordfish, etc. In the Baltic, where they have to travel some distance before reaching the open waters of the North Sea, Eels have been marked and recaptured; one such had travelled nearly 800 miles in 93 days. In the Straits of Messina the whirlpools bring to the surface a number of the inhabitants of the depths, and from November to July these include Silver Eels with eyes larger, colour darker, and sexual organs more developed than in those captured when migrating. Ripe Eels, nearly ready to spawn, are very seldom seen, and when taken in fresh water or in shallow water near the shore must be regarded as abnormal. Schmidt, a Danish zoologist, to whom we are indebted for the most recent and most complete account of the life-history of the Eel, describes and figures a ripe male Eel, 14 inches long, taken at the beginning of September 1903, in Præsto Fjord. This example is remarkable for the large size of the eye, the diameter of which is greater than the length of the snout. A ripe female Eel in the British Museum was taken at Mullingar, Westmeath, in the middle of Ireland, in October 1906; it is 27 inches long; the eye is not large for a Silver Eel, its diameter measuring only a little more than one-half the length of the snout and considerably less than the inter-ocular width, but the sexual organs are enormously developed and greatly distend the abdomen. Two very similar but smaller examples, captured at Toom Bridge, where the Bann leaves Lough Neagh, were described by J. T. Cunningham in 1896.

It seems pretty certain that all the Eels which go down to the sea from the British Isles spawn in the winter or spring in the Atlantic to the west and south-west of Ireland, at depths of 500 fathoms or more, and then die; at any rate, none of them ever come back. The eggs and earliest young have not yet been certainly identified, but it is probable that the former are large and float at or near the surface, where the latter also hatch out, feed, and grow.

A group of small transparent fishes, the Lepto-cephalids, were for a long time a puzzle to naturalists; in 1864 an American, Dr. Th. Gill, expressed the view that one of them, known to zoologists as *Leptocephalus morrissii*, was a young Conger, a conclusion which was arrived at independently a few years later by a Frenchman, Dareste, and in 1886 was proved to be true by another Frenchman,

Delage, who kept a specimen in an aquarium from February to September and saw the transparent ribbon-shaped larva become more opaque and more cylindrical and change into a little Conger. It was not until 1897 that the form known as *Leptocephalus brevirostris* was shown by the Italian naturalists, Grassi and Calandruccio, to be the larva of the Common Eel.

This Leptocephalid is about 3 inches long and has a comparatively deep, strongly compressed, transparent body and a small head ; the snout is pointed and the eye moderately large ; the small mouth is furnished with a few pointed teeth ; the vent is placed far back and the continuous vertical fins are confined to the posterior third of the body. Such larvæ (Pl. XXII, Fig. 1) have been taken in considerable numbers in the Atlantic Ocean to the west of the British Isles over deep water, 500 fathoms or more, from May to August, swimming near the surface at night and at some distance below the surface during the day ; the feeding, which has enabled the unknown younger fish to develop into a Leptocephalid, has now ceased and the digestive tract is empty. From September to November the Leptocephalids of the Eel may still be captured in the same place, but the metamorphosis has commenced, the body is less deep, the larval teeth have disappeared, the snout is more rounded, the eye is smaller, and the dorsal and anal fins have begun to grow forward (Pl. XXII, Fig. 2). From November onwards the migration towards the coasts and rivers is in progress and the transformation continues (Pl. XXII, Figs. 3 and 4), accompanied now by a decrease in length as well as in height ; finally, the

PLATE XXII

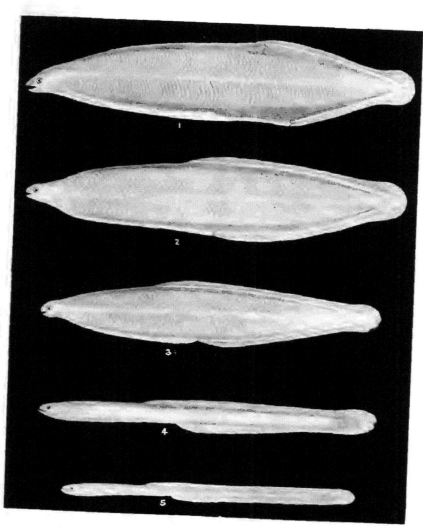

TRANSITION FROM LEPTOCEPHALID (1) TO ELVER (5)

body becomes pigmented, teeth are developed, and the larva has changed into a young Eel, or Elver, about $2\frac{1}{2}$ inches long (Pl. XXII, Fig. 5), which commences to feed and grow.

As a rule, the nearer the coasts or rivers are to the spawning-places the earlier do the elvers arrive, the greater are their numbers, and the less advanced is their development; thus some rivers are reached before the metamorphosis is completed, and the shrinkage continues in fresh water. The elvers commence ascending some of the rivers in the west of Ireland in the beginning of the winter, but do not appear to reach the Tay until April or May. However, there are other factors besides geographical position which determine the period of ascent; these are not yet understood, but in the case of rivers which are late, but from their situation should be early, it is probable that the elvers spend some weeks off the river mouths before entering.

As Eels of all sizes are found in the sea round our coasts it is evident that a good many do not enter the rivers as elvers, and it may well be the case that some may migrate into fresh water when they have grown into fair-sized Eels. In the spring and early summer the elvers enter many of our rivers in enormous numbers, forming dense columns, and wriggling over weirs or other obstacles to their progress in swarms; this migration of the elvers is in many places known as an Eel-fare.

The elvers are often captured and used for food, but in many places it is the custom to assist them in their journey in order that the future supply of large Eels shall not fail. According to Yarrell, on the Severn they were taken in great quantities with

12

sieves of haircloth, or even with a common basket, and were fried in cakes or stewed and considered delicious, whilst Thompson wrote of the Bann that in the early summer the river was black with thousands of young Eels 3 or 4 inches long, and that hay-ropes were suspended over the rocky parts to help them in their ascent.

As has already been mentioned, our knowledge of the life-history of the Eel has been greatly increased by the work of two Danes, Petersen and Schmidt. It is to a third Danish zoologist, Gemzöe, that we are indebted for the determination of the age and rate of growth of the Eel. He has shown that the elvers, which reach Denmark about May, grow slowly and are less than 4 inches long after a year spent in fresh water; during the next summer they attain a length of about 5 (4 to 7) inches; in the third summer this length is increased by about 4 inches, and it is in this season, when they are a little more than 7 inches long, that scale formation begins.

When a scale is examined under the microscope the outer surface is seen to be studded with little calcareous buttons, which are arranged in zones or rings parallel to the edges, and are separated from each other by narrow rings occupied only by the fibrous ground-substance of the scale. This structure is due to the fact that the Eel feeds and grows actively in the summer months only, and the zones are annual rings formed during the summer, whilst the narrow interspaces represent the growth of the scale during the colder months.

Scales are first formed on the middle of the side, and if scales from this region be examined their age in years is ascertained by counting the complete zones; the age of the Eel is that of the scale, with the addition of three or four years, one or two occupied in reaching the elver stage, two for the time passed before the young Eel began to form scales; in the autumn only half a year should be reckoned for the last zone, *i.e.* for migrating Silver Eels half a year less should be added to the number of zones.

By applying this method Gemzöe has shown

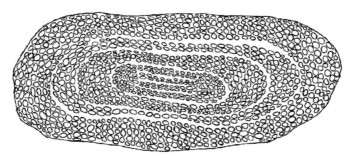

FIG. 18.—Scale of a Silver Eel showing five annual rings; the scale 4½, and the Eel from which it was taken 7½ or 8½ years old (*after Gemzöe*).

that the difference in size between the sexes only becomes apparent when the Eels are about six years old, when the females begin to grow much more strongly than the males. Male Eels were found to assume their breeding dress as a rule after spending five and a half or six and a half years in fresh water, the time varying from four and a half to eight and a half years; these were from 12 to 20 inches long. Female Silver Eels had usually spent six and a half to eight and a half years in fresh water, and measured from about 14 to 26 inches, whilst a few larger ones, about 3 feet long, were ascertained to have lived ten and a half to twelve and a half years since the elver stage. From this we may infer that some of the very large examples mentioned above may well have been twenty or thirty years old.

The different rate of growth of the male and female Eels affects the structure of the scales ; as a male fish is usually older and has grown more slowly than a female of his own size, his scales will have narrower and more numerous rings.

In addition to the two species of the North Atlantic there are several kinds of Eels inhabiting the Indo-Pacific from the Cape of Good Hope to the Sandwich Islands, but none are found on the Pacific coast of North America, nor in South America, nor in West Africa south of Morocco

As Schmidt has lately shown, the extraordinary life-history of the Eel furnishes the clue to this curious distribution of the species. He has found that at their breeding-places the water never has a temperature of less than $7°$ centigrade, and that they are only found on coasts and in rivers which are within reach of suitable breeding-places ; thus their absence from South America and West Africa is due to the coldness of the depths of the South Atlantic.

The Black Sea contains water of the right depth and temperature for the propagation of Eels, yet none are found in it or in the rivers connected with it ; but the salinity is here much lower than in the ocean and the deep water contains sulphuretted hydrogen in such quantities as to exclude all higher organic life, so that the absence of Eels is not surprising.

Readers who are interested in the biology of the Eel should consult the following recent memoirs, the last two of which contain references to the literature of the subject :—

(1) C. G. J. Petersen, " The Common Eel gets a Particular Breeding Dress before its Emigration to the Sea," in *Report of the Danish Biological Station for 1894.* Copenhagen (1896).

(2) K. J. Gemzöe, " Age and Rate of Growth of the Eel," in *Report of the Danish Biological Station for 1906.* Copenhagen (1908).

(3) J. Schmidt, " Contributions to the Life-History of the Eel," in *Rapp. du Conseil Internat pour l'Exploration de la Mer*, v. Copenhagen (1906).

(4) J. Schmidt, "On the Distribution of the Freshwater Eels throughout the World," in *Medd. fr. Kommission. f. Havundersogels Fiskeri*, iii. Copenhagen (1909).

The flesh of the Eel is rich and nutritious, but sometimes considered indigestible; Silver Eels are generally preferred to Yellow Eels for the table In some countries, Scotland for example, there is a prejudice against Eels as food, but in most places where they occur the fisheries are highly valued.

CHAPTER X

THE CYPRINOIDS

The Ostariophysi. The Carp family : synopsis of the British species—the Carp—Carp × Crucian Carp—the Crucian Carp—the Gold-fish—the Barbel—the Gudgeon—the Tench—the Minnow—the Chub—the Dace—the Roach—the Rudd—Roach × Rudd—White Bream—Common Bream—Bream × Roach—Bream × Rudd—White Bream × Rudd—the Bleak—Bleak × Roach—Bleak × Chub. The Loach family : the Loach—the Spined Loach

THE Carp family (*Cyprinidæ*) and the Loaches (*Cobitidæ*) represent in our waters the important order *Ostariophysi*, which includes a large proportion of the freshwater fishes of the world. Some of them, such as the Cat-fishes and the Electric Eels, are strikingly different in appearance from the Carp tribe, but all these diverse types agree in the possession of a very peculiar apparatus, formed by the modification of the four anterior vertebræ, some of the elements of which ossify separately and form a chain of little bones on each side—the Weberian ossicles — which connect the air-bladder with the ear. The two first of these represent the neural arch of the first vertebra, and form the wall of a backward prolongation of the perilymph spaces surrounding the internal ear ;

they are connected by ligament with the *tripus*, which is the modified rib of the third vertebra, inserted in the wall of the air-bladder, and in this connecting ligament appears the *intercalarium*, which is the neurapophysis of the second vertebra. The functions of this mechanism are not well ascertained, but it is probably the case that by its means the perception of sounds, movements, or alterations of pressure is intensified.

The Carp family (*Cyprinidæ*) is probably richer in species than any other family of fishes, considerably more than one thousand having been described from the lakes and rivers of Europe, Asia, Africa, and North America. They are fishes with a naked head, and usually a scaly body, and with all or most of the fin-rays jointed and flexible. The pectoral fins are placed low, the pelvic fins are abdominal in position, and there is no adipose fin. The mouth is toothless, more or less protractile, bordered above by the præmaxillaries only, and the gill-membranes are usually joined to the isthmus, restricting the gill - openings from below. Most characteristic of the family are the falciform lower pharyngeal bones, bearing a small number of teeth which bite upwards against a hard plate attached to a backwardly directed process of the basal part of the skull; the form of these pharyngeal teeth, their number and their arrangement in one, two, or three series are of great importance in the distinction of genera and species.

All our Cyprinoids are more or less gregarious ; they usually subsist on a mixed diet, but some, like the Carp, are almost exclusively vegetarian, whilst others, such as the Chub, are more predaceous.   All spawn in the spring or early summer, and apparently but once a year, although the spawning season may in some species be prolonged, or may be retarded from various causes   At this season even the most sluggish species become lively and sport at the surface of the water; the fish usually crowd together on quiet shallows, and seem to forget everything except the occupation of the time, so that they may readily be captured, and fall victims to the carnivorous fish and birds, which find them an easy prey.   The breeding males are sometimes remarkable for their brilliant coloration; usually they acquire little tubercles on the head, which in some species extend on to the back and sides of the body; another characteristic male feature is a thickening of the first ray of the pectoral fin.

After spawning, many of our Cyprinoids repair to rapid shallows to recuperate, the more sluggish species forming an exception to this rule; thence they make their way to the feeding-grounds adapted to the peculiarities of the species; in the winter all retire into deep water; some fall into a sort of torpor and lie huddled together in shoals, or may even, like the Tench, burrow into the mud; others, like the Roach and Chub, continue feeding whenever anything eatable is to be found.

The following synopsis, based on external characters only, indicates the arrangement of the British species adopted in this work, and may be used as a key for their identification :—

I. Dorsal fin long, with 14 to 23 branched rays; anal fin short; last simple ray of each more or less distinctly spinous and serrated.

    A. Two barbels on each side of the mouth; caudal fin strongly emarginate . 1. *Carp*

    B. No barbels; caudal fin slightly or moderately emarginate.

Dorsal spine feeble, finely serrated; anterior branched rays increasing in length backwards, the edge of the fin convex; $6\frac{1}{2}$ to 9 scales between origin of dorsal fin and lateral line . . . . 2. *Crucian Carp*

Dorsal spine strong and coarsely serrated; anterior branched rays the longest, the edge of the fin straight or concave; 5 to $6\frac{1}{2}$ scales between origin of dorsal fin and lateral line . . . 3. *Gold-fish*

II. Dorsal and anal fins short or of moderate length, neither with more than 13 branched rays.

    A. Mouth with barbels.

Mouth inferior; two barbels on each side; last simple ray of dorsal fin a serrated spine; 52 to 70 scales in the lateral line . . . . . 4. *Barbel*

Mouth inferior; one barbel on each side; no dorsal spine; 39 to 45 scales in the lateral line . . . . 5. *Gudgeon*

Mouth terminal; a small barbel on each side; scales small . . . . 6. *Tench*

    B. No barbels.

        1. Scales small; each tubule of the lateral line extending the whole length of the exposed part of the scale . . . . . . 7. *Minnow*

2. Scales moderately large; tubules of the lateral line not extending to the free edges of the scales.

    a. Free edge of anal fin convex; cleft of mouth nearly or quite reaching to below the anterior margin of eye . . . . . 8. *Chub*

    β. Free edge of anal fin concave. Origin of dorsal fin above or slightly behind base of pelvic fins.

Greatest depth of the body contained 3 to 5 times in the length to the base of the caudal fin; dorsal fin with 7 or 8 branched rays . 9. *Dace*

Greatest depth of the body contained $2\frac{1}{2}$ to 4 times in the length to the base of the caudal fin; dorsal fin with 9 to 11 branched rays . 10. *Roach*

Origin of dorsal fin far behind the base of the pelvic fins . . . . 11. *Rudd*

III. Dorsal fin short or of moderate length; anal fin long, with 15 to 29 branched rays; abdomen behind the pelvic fins compressed to a sharp edge over which the scales do not pass.

A. Body deep.

44 to 50 scales in the lateral line, 8 to 11 between origin of dorsal fin and lateral line; dorsal fin usually with 8, anal fin with 19 to 24 branched rays . . . . . 12. *White Bream*

49 to 57 scales in the lateral line, 11 to
15 between origin of dorsal fin and
lateral line; dorsal fin usually with 9,
anal fin with 23 to 29 branched
rays . . . . 13. *Bream*
B. Body elongate; anal fin with 15 to 20
branched rays . . . . . 14. *Bleak*

Three species of Carp are found in our islands;
they are readily recognized by the long dorsal and
short anal fins, in each of which the last simple ray
is more or less distinctly spinous, and has the
posterior edge serrated. Of these three species, two,
the Common Carp and the Gold-fish, are natives of
China, and have been introduced into various parts
of the world; the third, the Crucian Carp, is a
European species, but perhaps not indigenous to
Britain.

THE COMMON CARP (*Cyprinus carpio*) is placed
in a genus distinct from that of the Crucian Carp
and Gold-fish (*Carassius*) on account of the presence
of two barbels on each side of the mouth, and because the molariform pharyngeal teeth are arranged
in three series on each side; the main row is composed of three teeth, and in addition there are two
small anterior teeth placed one in front of the other,
whereas in the Crucian Carp and Gold-fish there is
a single series of four teeth on each side.

In the Carp the dorsal fin is highest anteriorly,
its spine is fairly strong and distinctly serrated,
whilst the branched rays are seventeen to twenty-two
in number; the caudal fin is deeply emarginate.
There are thirty-four to forty scales in the lateral

line, and five to seven in a transverse series from the origin of the dorsal fin to the lateral line. The colour is usually of a greenish brown, with brassy reflections on the sides.

According to continental writers the Carp may attain a weight of nearly 100 lbs., but 25 lbs. seems to be about the limit for this country; our figure (Pl. XXIII, Fig. 1) is of quite a small fish, only 7 inches long.

The date of the introduction of this species is uncertain, but it is now widely distributed in Britain and Ireland. The first-known mention of it as a British species is in Dame Juliana Berner's *Boke of St. Albans*, published in 1496.

The Carp is very variable in form and in the size of the fins; of domesticated varieties bred on the Continent may be mentioned the Mirror-carp, with the scales much enlarged, and usually reduced to one or two series on each side, the rest of the fish being naked, and the Leather-carp, entirely scaleless, and with the skin thickened.

Carp are chiefly found in lakes, ponds, and slow-running rivers; they are especially partial to weedy places where the bottom is muddy. In the summer-time the shoals of Carp may be seen lying motionless or slowly swimming about near the surface of the water, sometimes browsing on the weeds; or they may be feeding at the bottom, taking in and then ejecting mouthfuls of mud, from which they probably extract some nutriment in the form of decomposed vegetable matter; their diet is principally vegetarian, but worms, shrimps, insects, etc., are also eaten. In the winter the Carp retire into deep water and cease feeding; they have been

PLATE XXIII

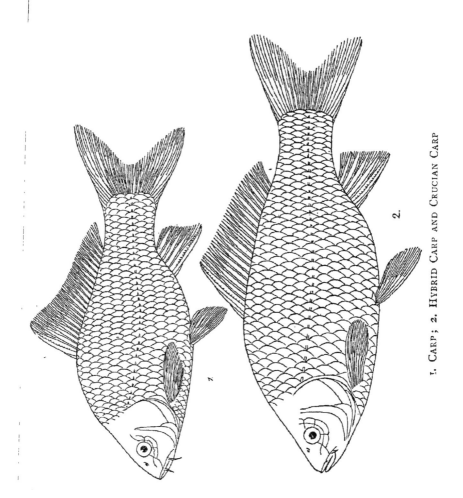

1, CARP; 2, HYBRID CARP AND CRUCIAN CARP

described as lying immovable on the mud, with their heads together, forming a circle.

They are extremely cautious, and it is a problem for the angler to find a line strong enough to hold Carp and yet fine enough to escape their notice, whilst the net fisherman finds them full of ruses to escape him, such as burrowing under or leaping over the edge of his seine.

Writing of the Carp in the Norfolk Broads, the Rev. R. Lubbock says: " Mrs. Glasse's receipt for stewing these fish begins with these words, ' First catch your Carp.' The worthy dame here displays acumen. In our extensive waters no fish baffle the fisherman so completely as large Carp. Every one has his peculiar tale of disappointment—how he surrounded a shoal of Carp with his turning net, and some sprang over and the rest ' mudded '; or how he found them working in a dyke, and placed a trammel net above and below them, so as to cut off retreat both ways, and then dragged the intermediate space with a third net, and got only one of the smallest. I believe that some of our fen-men regard this fish with mysterious awe; his exits and entrances puzzle them—they regard him as something more than a fish, and look upon him as what the Scotch call ' No cannie.'"

Carp can live for a long time out of water, and can be sent alive for considerable distances packed in damp moss; they often breathe atmospheric air under normal conditions in the warm weather, coming to the surface and taking it in with a noise which has been described as "smacking their lips."

The breeding season is usually in May or June, but may be prolonged throughout the summer, especially if the weather be cold; at this time the males have little white tubercles on the cheeks and opercles. The Carp spawn on quiet shallows amongst the weeds, to which the eggs adhere. Sometimes a female is attended by two or even three males, which swim above her, often leaping out of the water, and by other antics betraying their excitement.

The eggs are small and very numerous; they hatch out in about a fortnight or sometimes less, and the fry grow rapidly; sexual maturity may be attained in three years, when the fish may be a foot in length and weigh about 1 lb., but the rate of growth varies enormously according to circumstances, and fish of this age may be three or four times as large; growth takes place in the summer months only.

The Carp is popularly considered to rival the Pike in longevity, and has been said to attain an age of a hundred and fifty years; but it must be confessed that this and similar statements rest on very unreliable evidence, although there is reason for supposing that Carp live to a good old age.

It often happens that apparently wildly improbable tales may have far more truth in them than others which seem much less fanciful, and this may well be the case with the following passage from Walton :—

" I have both read it, and been told by a gentleman of tried honesty, that he has known sixty or more large Carps put into several ponds near to a house, where, by reason of the stakes in the ponds, and the owner's constant being near to them, it was impossible they should be stole away from him; and that when he has, after three or four years, emptied the pond, and expected an increase from them by breeding young ones (for that they might do so, he had, as the rule is, put in three melters for one spawner), he has, I say, after three or four years, found neither a young nor old Carp remaining And the like I have known of one that had almost watched the pond, and at a like distance of time, at the fishing of the pond, found, of seventy or eighty large Carps, not above five or six; and that he had forborne longer to fish the said pond, but that he saw, in a hot day in summer, a large Carp swim near the top of the water with a frog upon his head; and that he, upon that occasion, caused his pond to be let dry; and I say, of seventy or eighty Carps, only found five or six in the said pond, and those very sick and lean, and with every one a frog sticking so fast on the head of the said Carps, that the frog would not be got off without extreme force or killing."

We find this account confirmed by many modern instances, which are thus summed up by Smitt: "Carp are often troubled by the male frogs, which under the influence of sexual excitement attach themselves firmly to the head of the sluggish Carp, and with their forefeet press the eyes of the fish so hard as to produce blindness."

Carp are fairly good eating when well cooked, but all the recipes seem to agree in recommending the use of plenty of wine and other accessories which disguise the muddy flavour of the fish. They are easily kept in ponds, and are largely cultivated on the Continent, but not much in this country.

13

The name Carp is represented by similar words in many languages, such as the German *Karpf*, the French *Carpe*, the Celtic *Cerpyn*, and the Latin *Cyprinus* or *Cyprianus*, probably derived from Cyprus, the abode of Venus, in allusion to the fecundity of some fish of this family.

The Common Carp and the Crucian Carp form a hybrid, which is represented in the British Museum Collection by several examples, including a fine specimen, 16 inches long, from Norwich; one of half that length is shown on Pl. XXIII, Fig. 2.

This hybrid differs from the Carp especially in that the posterior barbels are quite short and the anterior ones very small or absent, whilst the dorsal spine is more finely serrated and the caudal fin is less deeply emarginate. The pharyngeal teeth are variable, forming one to three series, often two, with four teeth in the inner and one in the outer row. The branched rays in the dorsal fin number seventeen to twenty, the scales in the lateral line thirty-three to thirty-eight.

THE CRUCIAN CARP (*Carassius carassius* or *C. vulgaris*) differs from the Common Carp in that barbels are absent and the pharyngeal teeth are somewhat compressed and are arranged in a single series. The dorsal fin is more elevated, the branched rays, fourteen to twenty-one in number, at first increasing in length, sometimes nearly to the middle of the fin, whilst the spine is slender and feebly serrated; the caudal fin is but slightly emarginate. The body is usually deep, but varies considerably, and is sometimes as elongate as in the Carp, the name Prussian Carp being applied to fish of this

Plate XXIV

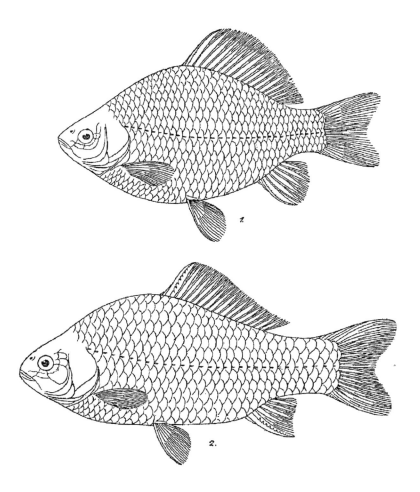

1. Crucian Carp ; 2. Gold-fish

form.   The scales in the lateral line number twenty-eight to thirty-five, and there are six and a half to nine between the origin of the dorsal fin and the lateral line.   The coloration is similar to that of the Carp.

The Crucian Carp attains a length of about 18 inches and a weight of about 7 lbs.; the example figured (Pl. XXIV, Fig. 1) is a small fish of 6½ inches.   This species is found throughout Europe and in Turkestan, Siberia, and Mongolia. In our islands it is by no means so common as the Carp, and it may not be indigenous; it seems to be absent from Scotland, Wales, and Ireland, and except in the Thames system and some of the Eastern counties is rare in England.

In habits the Crucian Carp closely resembles the Common Carp, but it is even more sluggish and more shy; on hot summer days the shoals may bask or swim about near the surface, but they usually lie quietly in the mud or among the weeds.   The breeding season is in April, May, and June, when the eggs are shed amongst the weeds.

The word Crucian corresponds to the German name for this fish, *Karausche*, which has been latinized to form the generic name *Carassius*.

THE GOLD-FISH (*Carassius auratus*) is closely related to the Crucian Carp, but the body is usually more elongate, the dorsal fin is highest anteriorly, and has the spine strong and coarsely serrated, and the scales are larger, numbering twenty-five to thirty in the lateral line, and five to six and a half between the origin of the dorsal fin and the lateral line.   Like the Common Carp, this species is a native of Eastern Asia, and in a wild state has the greenish-brown hue of the other Carps, and it is only the domesticated varieties which lose their black or brown pigment and exhibit the well-known gold and silver coloration. Some of the largest specimens I have seen, certainly several pounds in weight, were in the ponds at Hampton Court. A small example, 7 inches long, is shown on Pl. XXIV, Fig. 2.

The Chinese and Japanese have produced many remarkable varieties of Gold-fish; the dorsal fin may be variously reduced, and the caudal may split vertically into its component halves, thus showing four lobes instead of two, or if the upper edges do not separate it may appear trilobed. The most curious type is the Telescope Fish, with protruding eyes, no dorsal fin, and a very large trilobed caudal.

THE BARBEL (*Barbus barbus* or *B. vulgaris*) belongs to a genus which is very rich in species; nearly two hundred different forms are known from Africa alone, and numerous kinds are also found in the Indian region, including the famous Mahseer (*B. mosal*), so well known as a sporting fish, which attains a weight of more than 100 lbs.

The European species are few in number and our Barbel is the only one found north of the Alps; its range on the Continent extends from France through Germany to the Danube, but it is absent from the northern parts of Europe. In the British Isles it seems to be confined to the Thames, the Trent, and some of the Yorkshire rivers.

The Barbel has a rather elongate body covered with scales of moderate size, numbering fifty-two to seventy in a longitudinal series. The head is noticeable for the high position of the smallish eye, the rather long snout, with the upper profile more or less decurved, and the inferior horseshoe-shaped mouth with thick lips and with two barbels on each side, the anterior near the end of the snout, the posterior near the corner of the mouth. The dorsal fin is placed in the middle of the length of the fish; the last simple ray is a strong serrated spine, and the branched rays number eight or nine, whilst the anal has only five. The pharyngeal teeth are sub-conical, hooked, arranged in three series on each side, five in the inner row, three in the middle, and two in the outer row. The general colour is a greenish olive, darkest on the back, and with golden reflections on the sides; small dark brownish spots are often present on the back and sides and on the dorsal and caudal fins.

In England the Barbel has been known to attain a length of more than 3 feet and a weight of about 20 lbs., but much larger specimens have been recorded from the Danube. The largest English specimen I have seen weighed 12 lbs. 12 oz. and was 30 inches long, it was taken in the Kennett; an example from the Thames, 9 inches long, is shown on Pl. XXV, Fig. 1.

Barbel feed especially on worms, shrimps, insect larvæ, etc., but they appear not to disdain any sort of animal or even vegetable matter, which they find by rooting about on the bottom or in the banks with their snouts, often turning over stones and using their barbels as feelers to aid them in their search for food. In the winter-time they herd to-gether in large shoals and retire into deep water, where they lie in a torpid condition.

The spawning season is in May or June, when the males acquire little tubercles on the head and the anterior part of the back; the fish assemble in shallow or fairly deep water where the bottom is gravelly; the eggs are said to be covered over with gravel by the parent fish, and to hatch out in about a fortnight. After spawning the Barbel make for the swift shallows, where they may be seen rolling about on the gravel.

The Barbel is a strong and active, yet wary, fish, and affords fine sport to the angler; opinions differ as to its value as food, the flesh being white and firm, but rather coarse; the eggs are more or less poisonous, sometimes inducing violent purging and vomiting, and also weakening the heart so much that fainting may result; the poisonous secretion is sometimes absorbed by the flesh of the lower part of the fish, which may thus produce similar effects, and to be safe it is best to eat Barbel only in the late summer and autumn, and to remove the roe as soon as possible after the fish is caught.

The Barbel derives its title from *Barbellus*, the diminutive of *Barbus*, the Latin name given to this fish in allusion to its beard-like appendages or barbels.

THE GUDGEON (*Gobio gobio* or *G. fluviatilis*) is rather similar to the Barbel in general form, as well as in the shape of the head, the structure of the mouth, and the position of the fins. However, there is no anterior pair of barbels, the eye is rather large, the scales number only thirty-nine to forty-five in the lateral line, and the unbranched rays of the dorsal fin are slender and articulated. The dorsal

PLATE XXV

1. BARBEL; 2. TENCH

fin has from six to eight branched rays, the anal five to seven. The colour is brownish or greenish above, silvery or golden on the sides and below, often with small scattered brownish spots on the upper parts and a row of larger blackish ones along the middle of the side; there are series of small dark spots on the dorsal and caudal fins.

The main reason for placing this species in a genus distinct from that of the Barbel is the arrangement of the pharyngeal teeth in only two series,

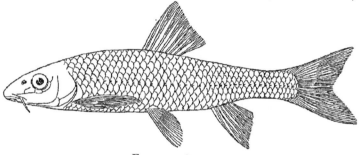

FIG. 19.—Gudgeon.

four or five in the inner and two or three in the outer on each side.

The Gudgeon is found all over Europe except the Iberian Peninsula and Greece, and extends through Russian Turkestan and Siberia to Mongolia. In England and Wales, it is generally distributed, but appears to be absent from the Lake District, Western Wales, and Cornwall; it is unknown in Scotland, but is common in Ireland.

This is a small species, rarely growing to more than 6 inches in length, whilst specimens of 8 inches are considered exceptionally large.

Gudgeon are usually found in rivers, especially

frequenting places where the bottom is gravelly or sandy and where the water is not very deep nor the current too strong; shoals of them may often be seen on the bottom quite close to the bank, lying still or moving about slowly in search of food, which they find by stirring up the sand or gravel and feeling with their barbels in much the same way as the Barbel; the food consists of shrimps, small molluscs, insect larvæ, worms, and the eggs and fry of other fish.

The breeding season commences in April or May, when the Gudgeons repair to the gravelly shallows for the purpose of spawning. At this time the males are ornamented with little tubercles on the head; the females deposit their eggs on the stones in small lumps, and one may take four weeks or more before all are shed, as all the fish are not ripe at the same time, the spawning may last far into the summer.

The fry hatch out in about a month, so that they may often be found at the same time as the breeding-fish.

Rusconi's account of the spawning of the Gudgeon in Lombardy has been thus translated by Smitt · "During my stay at Desio, on a most lovely day in July, I was walking early in the morning along the shore of the little lake of Villa Traversi Suddenly a noise reached my ear. I thought at first that some one was beating the water with sticks or with the flat of an oar. On glancing along the shore I soon detected the spot from which the sound proceeded, as well as the cause of the disturbance; it was caused by spawning fishes. Eager to obtain a closer view of this sight, I stealthily made my way towards them, and under cover of the bushes that fringed the shore I got near enough to observe them with care and without betraying my own presence. They lay at the mouth of a small brook, the water of which was cool and clear, but so scanty that the pebbles at the bottom were almost dry. They were Gudgeons. They approached the mouth of the brook. With rapid strokes they came swiftly on and advanced about a metre up the brook, not leaping, but in a manner gliding over the pebbles. After this first spurt they stopped, bent the trunk and tail alternately to the right and left, and in this way rubbed the ventral side against the bottom. With the exception of the belly and the lower part of the head their whole body now lay out of the water. They retained this position for seven or eight seconds. Then they dealt a sharp blow with the tail on the bottom, splashing the water in all directions, turned round, and darted back to the lake, soon to repeat the same operation."

The Gudgeon affords good sport in spite of its small size, as it readily takes a worm ; moreover, it is excellent food, of delicate flavour, and easily digested.

The name Gudgeon is derived from the Latin *Gobionem* through the French *Goujon*.

THE TENCH (*Tinca tinca* or *T. vulgaris*) is well distinguished from other British Cyprinoids. In form it is moderately deep and rather stout, with the dorsal profile convex from the head to the dorsal fin. The head is broad, with the eyes rather small, the snout blunt, and the mouth terminal and oblique, with a short barbel on each side near the corner.

The scales are quite small, numbering 99 to 120 in the lateral line. The dorsal and anal fins are rounded; the former originates above the base of the pelvics and has eight or nine branched rays, whilst the latter has from six to eight; the caudal fin is slightly emarginate, with rounded lobes. The coloration varies from greenish yellow to greenish black; sometimes a golden variety may be met with.

The Tench is found all over Europe and in Asia Minor and Western Siberia; it is generally distributed in the British Islands as far north as Loch Lomond, and is also found in the Tay and Dee systems, where it is probably not indigenous.

As a rule, the Tench does not exceed 8 lbs. in weight, but Yarrell mentions one of 11 lbs. 9 oz., and still larger examples have been recorded from continental localities; a specimen a foot long is figured (Pl. XXV, Fig. 2).

This is pre-eminently an inhabitant of still waters, but occurs also in slow-flowing rivers; it thrives in small weedy ponds with muddy bottom, in reservoirs, or even in clay-pits. It is a quiet and indolent fish, in the summer lying at the bottom, or sometimes, on calm, hot days, at the surface amongst the weeds, whilst it passes the winter buried in the mud in a sort of torpor. Siebold relates that some Tench which had thus buried themselves were dredged up and placed on the bank; they showed no signs of life, and after some time were awakened by several blows with a stick, when they regained the pond by a series of jumps. In the spring, when they emerge from their winter sleep, the Tench are lean, but they rapidly fatten on a mixed diet of mud, weeds, worms, insects, and little shellfish.

Like Carp, Tench can live for some time out of the water, and they have been known to live buried in the mud at the bottom of a pond which had dried up during a hot summer; unlike Carp, however, they are not particularly cunning, and readily fall victims to the angler who knows how and where to fish for them, whilst the method employed by the fishermen on the Norfolk Broads has been thus described by the Rev. R. Lubbock—

"Tench catching, as it is justly termed, originated with a family of the name of Hewitt, at Barton, all the members of which were fishermen and gunners. One of them, observing the sluggish nature of this fish, attempted to take them with his hands, and often succeeded. The art has spread, and the system is better understood, so that at this time there are in Norfolk fishermen who, upon *shallow waters*—for in deep nothing can be done thus—prefer their own hands, with a landing-net to be used occasionally, to bow-nets or any other engines. The day for this operation cannot be too calm or too hot. During the heat of summer, but especially at the time of spawning, Tench delight in lying near the surface of the water amongst beds of weeds; in such situations they are found in parties, varying from four or five to thirty in number. On the very near approach of a boat they strike away, dispersing in different directions, and then the sport of the tench-catcher begins. With an eye like a hawk, he perceives where some particular fish has stopped in his flight, which is seldom more than a few yards; his guide in this is the bubble which rises generally where the fish stops.

"Approaching the place as gently as possible in

his boat, which must be small, light, and at the
same time steady in her bearings, he keeps her
steady with his pole, and, lying down with his head
over the gunwale and his right arm bared to the
shoulder—taking advantage, in his search, of light
and shade—he gently with his fingers displaces the
weeds, and endeavours to descry the Tench in his
retreat.   If the fisherman can see part of the fish,
so as to determine which way the head lies, the
certainty of capture is much increased ; if he cannot,
immersing his arm, he feels slowly and cautiously
about until he touches it, which, if done gently on
head or body, is generally disregarded by this
sluggish and stupid fish ; but if the tail is the part
molested, a dash away again is the usual consequence.   Should the fisherman succeed in ascertain-
ing the position of the fish, which under favourable
circumstances he generally does, he insinuates one
hand, which alone is used, under it, just behind the
gills, and raises it gently, but yet rapidly, towards
the surface of the water.   In lifting it over the boat
side, which, it need not be said, should be low, he takes
care not to touch the gunwale with his kunckles,
as the very slightest jar makes the captive flounce
and struggle.   On being laid down, the Tench often
remains motionless for full a minute, and then
begins apparently to perceive the fraud practised
upon it.   The fisherman then, if he 'marked' more
than one Tench when the shoal dispersed, proceeds
to search for it.   If not, he endeavours to start
another, by striking his pole against the side or
bottom of the boat—several are generally close at
hand.   The concussion moves other fish, when the
same manœuvres are repeated.   In this way I have

seen fifteen or sixteen good-sized table Tench taken in a short space of time. And in the course of a favourable day one fisherman will easily secure five or six dozen."

The breeding season is usually in June, when the males may be distinguished by the greatly enlarged outer ray of the pelvic fin. The eggs are very small and numerous and are deposited on the weeds in quiet shallows, generally hatching out in about a week. The young fish usually attain a weight of about $\frac{1}{4}$ lb. in a year, and a fish of 8 lbs. will probably be about the same number of years old.

The Tench used to be popularly regarded as the physician of the fishes, especially of the Pike, and even now the name "doctor fish" is given to it in some parts of England. The sick or wounded fish were supposed to be cured by the touch of the Tench, the thick slime covering the body of the latter being said to act as a sort of balsam. This belief is now generally discredited, as is the idea that the Pike forbears to eat his physician out of gratitude for his services ; indeed, in some localities, a small Tench is said to be a very good bait for a Pike. My friend, the late Dr. Bowdler Sharpe, told me that one day in May he stood on the bridge over the lake at Avington and watched a large Tench lying in the water below; a shoal of Perch swam up and lay round and above the Tench and appeared to be rubbing against him ; on being disturbed they swam back under the bridge, but soon repaired again to the Tench and repeated this manœuvre several times. The meaning of this is obscure, but there can be little doubt that observation of similar

14

incidents has led to belief in the healing powers of the Tench, not only for other fishes, but for the human race.   In olden times it was considered a specific for many diseases, being applied to the hands and feet to cure fever, or laid over the region of the liver for jaundice; headache, toothache, and other ills were treated on similar lines.

The flesh of the Tench is white and firm, but varies much in quality, the best table fish sometimes coming from the foulest water.   It is said the muddy flavour is removed by scalding, thus getting rid of the coat of slime which covers the body.   In the Middle Ages this fish was much appreciated, and used to be kept by the monks in their stew-ponds.

The name Tench is derived through the old French *Tenche* from *Tinca*, the Latin name of the fish.

THE MINNOW (*Phoxinus phoxinus* or *P. aphya*) is very similar to the Chub and the Dace in form, in the structure of the mouth, the pharyngeal dentition, etc., but it differs in some important characters which entitle it to be placed in a distinct genus.   The scales are small, numbering from eighty to more than one hundred in a longitudinal series, and are scarcely imbricated, whilst the lateral line runs nearly along the middle of the side and is composed of scales whereon the tubules extend the whole length of the exposed parts; the lateral line is variously developed, in some examples not reaching beyond the level of the pelvic fins, in others extending nearly to the base of the caudal.   The dorsal fin is placed farther back than in the Dace or Chub, ending above the origin of the anal; both fins have from six to eight branched rays. The coloration is very variable, but is usually of a silvery grey, with the back green, brown, or nearly black, and with a golden band along the upper part of the side; sometimes a series of dark vertical bars descend from the back, sometimes there is a dark longitudinal band along the middle of the side, or there may be scattered irregular spots.

The Minnow is found all over Europe except the Iberian Peninsula, and in Russian Turkestan and Siberia; in Britain it ranges north to the River Deveron in Banffshire, and has been introduced into the Spey, but is absent from the Northern Highlands of Scotland; in Ireland it is local, and at one time was said to be restricted to the counties of Dublin and Wicklow. Major Trevelyan, however, has sent me some from Fermanagh. A closely allied form occurs in Spain and Portugal (*P. hispanicus*).

As a rule, the Minnow does not grow to a length of more than 3 or 4 inches, and one of the last-named size is shown on Pl. XXXII, Fig. 3; occasionally specimens 6 or 7 inches long have been taken.

These pretty little fish are found in lakes, ponds, canals, and rivers, but they especially prefer clear streams where the bottom is sandy or gravelly; they swim in companies, and in the larger rivers are often to be found where the water begins to deepen at the end of a shallow. Sometimes long processions may be seen moving from one place to another, in search of new feeding-grounds; on such occasions they are said implicitly to follow the leaders, whose daring or curiosity may not always be for the benefit of the shoal; in fact, Minnow nets and traps depend for their success on these characteristics of the species. It is only necessary to lower a net into the water where Minnows abound, and to let it remain still; they will soon gather round, and then the boldest will venture over it and will quickly be followed by the rest, especially if their inquisitiveness be stimulated by the presence of a few pieces of red wool or some other bright material; when the shoal are busily engaged in testing and discussing the nature of this new object the greater part of them may be captured if the net is quietly drawn up.

In the winter the Minnows retire into deep water and lie hidden under stones or in holes in the banks. Their food consists principally of insects, worms, shrimps, little molluscs, and the eggs and minute fry of other fishes. They are also said to eat dead fish, but only on the authority of a writer in *Loudon's Magazine*, quoted by Yarrell, who related that as he crossed a brook he "saw from the foot-bridge something at the bottom of the water which had the appearance of a flower. Observing it attentively I found that it consisted of a circular assemblage of Minnows ; their heads all met in the centre, and their tails diverging at equal distances, and being elevated above their heads, gave them the appearance of a flower half-blown. One was longer than the rest, and as often as a straggler came in sight, he quitted his place to pursue him ; and having driven him away, he returned to it again, no other Minnow offering to take it in his absence. This I saw him do several times. The object that had attracted them all was a dead Minnow, which they seemed to be devouring."

The spawning takes place in May and June, when the males are distinguished by the thickened pectoral fin-rays, the presence of little whitish tubercles on the head, which is otherwise blackish, and the scarlet colour of the belly. The Minnows repair in large numbers to gravelly shallows, usually in brooks where the stream runs fairly rapidly. Here the eggs are deposited on the bottom, and adhere to each other and to the stones, so that they are not washed away by the current. After the spawning the males have been observed to ascend the shallows in large shoals and to lie together, forming an almost solid mass, which looks like a bed of weeds, whilst the little white spots on their heads have been compared to half-open buds ; after some days they return and recommence an active life.

The chief use of the Minnow is as a bait for Trout, Perch, and Pike, but it is not to be despised as food ; indeed, according to Walton, a Minnow-tansy is a dainty dish of meat, and Day tells us that in 1394 seven gallons of Minnows were served at a banquet given by William of Wykeham.

The name Minnow appears in old writings in various forms such as Menoun, Minoe, etc. ; it seems to have some connection with the French *menu*, small, but philologists are not agreed as to its derivation.

THE CHUB (*Leuciscus cephalus*) has the body rather elongate and little compressed ; the greatest depth usually measures about one-fourth of the length to the base of the caudal fin, but in large specimens is sometimes not much less than one-third of that distance ; in the adult fish the width of the flat interorbital region is nearly one-half of the length of the head; the mouth is large, its cleft often reaching the vertical from the anterior edge of the eye. The pharyngeal teeth resemble those of the Roach in structure, but are arranged in two series, five in the inner and two in the outer on each side. There are forty-two to forty-nine scales in the lateral line, seven or eight in a transverse series from the origin of the dorsal fin to the lateral line, and three or four from the latter to the base of the pelvic fin. The branched rays in the dorsal and anal fins number seven to nine; the former originates a little behind the base of the pelvic fins, and has the free edge straight or a little convex, whilst the latter is always distinctly convex. The coloration is silvery, or in larger specimens coppery, with the back greenish or brownish; each scale is dark at the base; the dorsal, caudal, and pectoral fins have the colour of the back, and the pelvic and anal fins are usually reddish.

The Chub is found all over Europe except in the extreme north and the Iberian Peninsula; in the latter, however, it is represented by a closely related form; it also ranges through Asia Minor to Persia. In Britain it occurs in most rivers south of the Firth of Forth, except in West Wales, Devon, and Cornwall, but it is absent from Ireland.

In this country the Chub grows to a length of about 2 feet and a weight of 8 lbs., but in some of the larger rivers on the Continent a weight of 12 lbs. is attained. Two fine examples, 23 inches long, from the Hampshire Avon, were exhibited at the Sports Exhibition held in Vienna in 1910. One, weighing 7 lbs. 6½ oz., was caught by Mr. F. W. Smith; the other, a fish of 7 lbs. 5 oz., by Mr. E. J.

PLATE XXVI

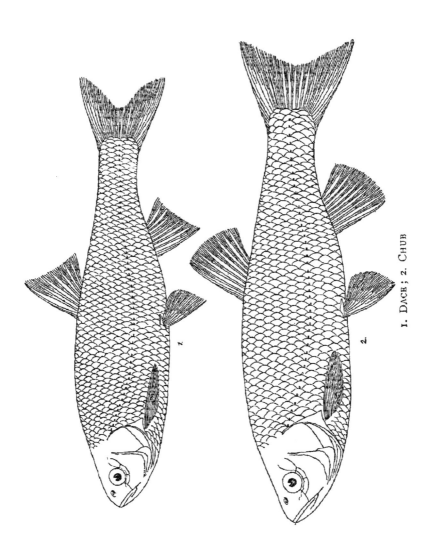

Walker; the example figured (Pl. XXVI, Fig. 2) is less than a foot long.

This is especially a river fish, and in hot summer days may often be seen in shoals lying near the top of the water and leaping at the flies, or just as often feeding near the bottom on weeds, shrimps, worms, etc., frequenting swift shallows such as mill-streams, from which habit it has in France received the name of *Meunier* (Miller).

The larger Chub are often solitary and are very wary, often lying under the shelter of an overhanging bank or beneath the shade of some bush; they are extremely voracious and feed to a considerable extent on little fish such as Minnows and Gudgeon, a prey for which their large mouth and strong jaws suit them. In the autumn the Chub retire into deep water and remain there throughout the winter.

The breeding season is from April to June, when the fish assemble on weedy shallows where the bottom is gravelly and the current moderately strong; here they may be seen for some days leaping out of the water and making a considerable disturbance. At this season the males have little tubercles on the head, and their scales also become rough. The eggs adhere to the stones and hatch out in a week or a little more; the fry are said to grow rapidly, and the fish to become mature at the age of three years.

As food the Chub is not of much account, being coarse and bony, but with some trouble may be made into a passably good dish.

The various names by which the Chub is known seem mostly to have reference to the size and breadth of the head; Chub is evidently from the same root as chubby, meaning with fat cheeks;

Chevin and Chavender are equivalent to the French *Chevaine*, probably from *chef*, a head; Loggerhead needs no explanation. The north-country name *Skelly* seems to be only a variant of *Schelly*, which is the name of the white-fish found in Ullswater and Haweswater; these names seem to have reference to the conspicuous scales.

THE DACE (*Leuciscus leuciscus*) is closely related to the Chub; it is a rather slender fish, the greatest depth of the body measuring from one-fifth to not much less than one-third of the length to the base of the caudal fin. The head is shorter, the inter-orbital region narrower, and the mouth smaller than in the Chub. The scales number forty-seven to fifty-four in the lateral line, eight or nine in a transverse series from the origin of the dorsal fin to the lateral line, four or five between the lateral line and the base of the pelvic fin. The dorsal fin has seven or eight branched rays, originates above the base of the pelvic fins, and has the free edge concave; the anal fin has seven to nine branched rays and has a concave margin, in striking contrast to the rounded anal fin of the Chub. The coloration is silvery white; the back, with the dorsal and caudal fins, is greenish or brownish; the lower fins are whitish, sometimes tinged with red.

The Dace is found all over Europe north of the Pyrenees and the Alps, and ranges throughout Siberia; it inhabits most of the English rivers and those of Wales, except in the west, but is absent from Scotland and Ireland.

This is a small species, rarely growing to much more than a foot long; specimens of 1 lb. weight are considered large, although as much as $1\frac{1}{2}$ lbs. is sometimes attained; the example figured (Pl. XXVI, Fig. 1) measures 9 inches.

The Dace is a graceful and lively fish, which delights in clear streams; it feeds on insects, shrimps, worms, etc., sometimes leaping at the flies on the surface of the water and sometimes keeping near the bottom; it spawns at about the same time as the Chub and often in the same places, and like the Chub retires to deep water for the winter.

The Dace affords excellent sport to the fly fisherman; as food it is distinctly better than the Chub, but the chief use of this species is as a bait for Pike, its bright coloration, activity, and tenacity of life making it especially suitable for this purpose.

The name Dace is from the old English Darse or Dart, applied to the fish from its darting movements, and appropriately, as any one will acknowledge who has watched them in a mill-stream and seen them make their way against the current by means of a succession of quick darts.

In Lancashire the Dace is called " Graining," and Yarrell believed that this was a distinct form, to which he gave the name *Leuciscus lancastriensis*; but the characters he used to diagnose this supposed species pertained only to the individual specimen he described, and I cannot see that the Dace of Lancashire differ in any way from those of other parts of England.

THE ROACH (*Rutilus rutilus*) has the body moderately deep and more or less compressed, the greatest depth in the adult fish measuring from one-fourth to two-fifths of the length to the base of the caudal fin, and usually more than the length of the head, which is one-fifth to one-fourth of the same distance.   The small mouth is terminal or nearly so, and has the cleft more or less oblique, sometimes nearly horizontal ; the pharyngeal teeth are compressed, hooked, entire, or feebly pectinated, arranged in a single series, five or six on each side. The scales are rather large, varying in number from forty to forty-six in the lateral line, whilst there are from seven to nine longitudinal series between the

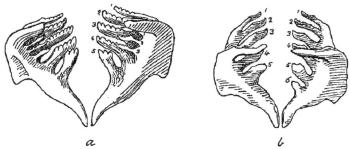

FIG. 20.—Lower pharyngeals of Rudd (*a*) and Roach (*b*), showing the difference in form and number of the teeth (*after Fatio*).

origin of the dorsal fin and the lateral line and three to four and a half from the latter to the base of the pelvic fins.

The dorsal fin originates above the end of base of the pelvic fins, is composed of three graduated simple rays and nine to eleven branched ones, is rather elevated, and has the free edge concave; the anal fin has three simple and nine to twelve branched rays, but its base is shorter than that of the dorsal.

The Roach is a silvery white fish, with the back greenish, and with the lower fins usually tinged with red; it is very variable in form, for some examples are as deep as the Rudd, others as slender as the Dace. The specimen figured (Pl. XXVII, Fig. 1) is 8 inches long, and is of average depth.

This species inhabits Europe north of the Pyrenees and the Alps, and in Asia it is found in Russian Turkestan and throughout Siberia. It does not occur in Ireland; in Devon, Cornwall, and West Wales it is scarce and local, whilst in Scotland its northern limits are Loch Lomond and the Teith.

Roach of more than 3 lbs. are extremely rare, and one does not often see finer specimens than two of 2 lbs. 4¾ oz. and 2 lbs. 5¾ oz., taken near Wilton in November, 1902, and presented to the British Museum by the Rev. H. G. Veitch. I am indebted to Mr. A. J. Alexander for particulars concerning a Roach of 3 lbs. 10½ oz., which I believe is the largest known from English waters. It was taken in November, 1904, from the Bristol Water Company's reservoir, made by damming a stream, and containing over 160 million gallons of water; except for Sticklebacks, this and about a dozen other fine Roach were the only fish in the reservoir, and the absence of enemies and competitors accounts to some extent for the large size attained. I have seen this fish, which is an undoubted Roach, and measures 17 inches in total length. The next largest I have seen, a fish of 2 lbs. 13½ oz., was captured in the Thames by Mr. G. Edmonds in November, 1903. It also is 17 inches long, but is not nearly so deep as the Bristol specimen.

Roach are found in lakes, canals, and in rivers which are not too rapid, preferring as a rule those which are slow and deep They swim in shoals, usually feeding near the bottom on weeds, insects, larvæ, little shellfish, etc., but on warm summer days sometimes taking flies at the surface. They are not particularly shy, except in waters which have been overfished, and from their abundance, and their readiness to take a bait, they probably give enjoyment to more anglers than does any other species of fish in our rivers. Nevertheless, it must not be supposed that the capture of Roach is an easy matter; the fish are so sharp-eyed that lines of only the finest gut must be used, or some anglers prefer a single horsehair next the hook; moreover, the fish have an irritating habit of taking the bait into the mouth and immediately spitting it out again, so that it is necessary to attain the art of perceiving the slightest movement of the float and of immediately making a corresponding movement of the wrist to strike the fish. I have often thrown in a pellet of paste, and have seen one Roach after another take it in and reject it, until one found it sufficiently to his liking to swallow it. Mr. Fennell, who watched Roach in the same way, declared that if the largest fish in the shoal had rejected the paste none of the others would touch it, and that after doing so he would often swim away elsewhere and the rest with him.

In the winter, when the weeds begin to decay, the Roach retire into deep water, but still feed on occasion. In April or May they band together in vast companies, and make their way to their breeding places, which are weedy shallows near the banks or in small tributary streams; here they press together in a dense mass, and by their movements against each other are said sometimes to produce a

PLATE XXVII

1. ROACH; 2. HYBRID ROACH AND RUDD; 3. RUDD

sort of hissing noise. In Norfolk they have been described as crowding together along the rushes which fringe the banks in such dense multitudes that every instant one may see small ones raised half out of the water by the passage of larger fish. The males are now quite rough, little conical tubercles being developed not only on the head, but all over the body.

The eggs, which are small and very numerous, are shed on the bottom, and are hatched in a fortnight or a little less; in eight to ten days the yolk-sac is absorbed and the young fry swim about in dense shoals among the reeds near the banks; they are said to attain sexual maturity when quite small, at an age of two years, whilst three-year-old fishes are said to be 4 or 5 inches long.

The flesh of the Roach is white and firm, but usually rather muddy; the best flavoured are those taken in clear, running water.

The name Roach is derived from the old French *Roche*, of uncertain origin and meaning.

THE RUDD (*Scardinius erythrophthalmus*) is usually a deeper fish than the Roach, the greatest depth of the body in the adult measuring not less than one-third and sometimes nearly one-half of the length to the base of the caudal. The terminal mouth is somewhat larger and more oblique than in the Roach, and the pharyngeal teeth differ notably from those of that species both in structure and number, as they are strongly pectinated and are arranged in two series, five in the posterior and three in the anterior row (Fig. 20, p. 220).

The scales number thirty-nine to forty-four in the lateral line, seven or eight in a transverse series from

15

the origin of the dorsal fin to the lateral line, and three or four between the latter and the base of the pelvic fin.

The dorsal fin has eight to ten branched rays; it is rather smaller and has the free edge less concave than in the Roach; also it is placed much farther back, originating well behind the base of the pelvic fins and ending nearly above the vent; the anal has ten to thirteen branched rays, and usually has a longer base than the dorsal.

It is worth noticing that in all the principal characters in which it departs from the Roach type, such as the double series of pharyngeal teeth, the backward position of the dorsal fin, and the elongation of the base of the anal, the Rudd approaches the White Bream, and there is still another Bream-like feature to be mentioned, that the belly behind the pelvic fins is compressed and keeled; however, the scaling is continuous over the keel, which is not the case in the Breams. In colour the Rudd differs from the Roach chiefly in having a bronze or golden tinge on the sides; the fins are usually reddish.

The Rudd is an inhabitant of Europe (except the Iberian Peninsula), Asia Minor, Russian Turkestan, and Siberia. In England and Wales it is rather local; it is common in the Norfolk Broads, but seems to be absent from the Trent. It occurs in a few ponds in Sussex, in the Isle of Wight, and in Slapton Ley, in Devonshire, but I am not acquainted with other localities in England south of the Thames and Severn. Apparently it is not found in Scotland, but in Ireland it is abundant throughout the island

The Rudd attains a length of about 18 inches and a weight of at least $3\frac{1}{2}$ lbs.; the example figured (Pl. XXVII, Fig. 3) is 9 inches long.

PLATE XXVIII

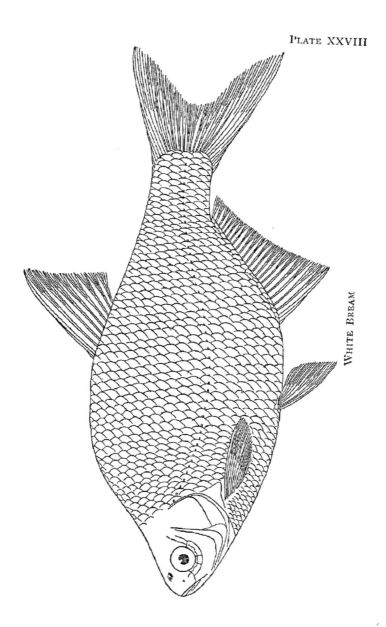

WHITE BREAM

This is especially an inhabitant of lakes and sluggish rivers, and even more than the Roach shows a preference for weedy places where the water is still. As a rule it is a quiet, inactive fish, feeding on the bottom like the Roach, but on warm summer days it often rises at the flies on the water. It is of a sociable disposition, and small companies will often attach themselves to shoals of Bream or Roach; it spawns at about the same time as the Roach, repairing to weedy shallows for the purpose ; at this season the shoals of Rudd make a characteristic noise by pouting, emitting air-bubbles at the surface, which float on the water and then burst.

The Rudd is not of much account as food, and except on the Norfolk Broads it does not attract the angler to any great extent. It takes its name probably from the red hue of the fins or from its general coppery tint, whilst the alternative "Red-eye" refers to the colour of the iris.

The so-called "Azurine" or "Blue Roach" of Knowsley, in Lancashire, was described by Yarrell, under the name *Leuciscus cæruleus*, as a silvery fish, with bluish back and pale fins. His specimens are in the British Museum and are quite small ; they do not seem to differ in any way from Rudd of the same size from other localities.

The hybrid between the Roach and the Rudd has been described from Bavaria, France, and Belgium, but has not, I believe, hitherto been recognized in this country. It is probably not very rare, but may often have been mistaken for one of the parent species. There are two examples of this hybrid in the collection of the British Museum, a fine specimen 11 inches long, from Thetford, and a smaller fish of $8\frac{1}{2}$ inches, from the Cam at Newport; the latter is shown on Pl. XXVII, Fig. 2.

These are in every way intermediate between the parent forms; the dorsal fin is rather more elevated and has the free edge a little more concave than is usual in the Rudd; it is also placed farther forward than in that species, but not so far forward as in the Roach, its origin falling in or a little behind the vertical through the tip of the pelvic axillary scale. The mouth is terminal, but less oblique than in the Rudd; resemblances to the latter are shown in the presence of an abdominal keel behind the pelvic fins and in the pectination of the pharyngeal teeth, but the number of the latter points unmistakably to the Roach parentage, there being six on one side and five on the other in the inner row, whilst the outer row is represented by a single tooth, or in the larger fish on one side by none. In both specimens I count nine branched rays in the dorsal fin and eleven in the anal, whilst the scales in the lateral line number forty-two or forty-three.

THE WHITE BREAM or SILVER BREAM (*Blicca bjoernka*) has a deep and strongly compressed body, with the abdomen behind the pelvic fins compressed to an edge, dividing the scales of the two sides. The snout is short and blunt, the mouth nearly terminal, with the cleft slightly oblique. The eye is rather large, its diameter never much less than the length of the snout. The pharyngeal teeth are in two series, five or six in the inner and two or three in the outer on each side.

The dorsal fin is rather elevated and pointed, and

is composed of three simple and eight, exceptionally seven or nine, branched rays; the branched rays in the anal fin number nineteen to twenty-four, and the lower lobe of the caudal is a little longer than the upper. The scales number forty-four to fifty in the lateral line, eight to eleven in a transverse series from the origin of the dorsal fin to the lateral line, and five to six and a half between the lateral line and the base of the pelvic fin. The coloration is silvery white, with the back greenish and the fins greyish.

The White Bream inhabits Europe north of the Alps and Pyrenees and extends into Western Siberia; in our country it appears to be confined to eastward rivers from Yorkshire to Suffolk.

This species attains a length of about a foot and a weight of $1\frac{1}{4}$ lbs. The example figured (Pl. XXVIII), 9 inches long, is from the Cam.

This species is found only in lakes or in slow-running rivers, where it swims in shoals, often in company with the Roach or Rudd. It feeds on or near the bottom, eating weeds, insects, worms, little shellfish, etc., and seems to have a curious habit of swimming upwards after seizing food, so that the angler becomes aware that one of these fish has taken his bait by the float, relieved of the weight of the shot, lying flat on the surface of the water. During the summer the White Bream frequents shallows, or places where the water is of moderate depth, but in the winter it retires into deep water.

The spawning usually takes place in May, when the fish assemble on shallows near the shore and splash about at the surface; the eggs are deposited on the weeds, to which they adhere.

The flesh of the White Bream is of poor quality, and as these fish are always lean and bony, and never attain any size, they have practically no value as food. They show little sport when hooked, and I have known them come straight out of the water as though they were dead, without even a wag of the tail; their greediness sometimes makes them rather a nuisance to the angler, who would prefer catching fish of more value.

THE COMMON BREAM or CARP BREAM (*Abramis brama*) has a more projecting snout, a more inferior mouth, and a smaller eye than the White Bream. The pharyngeal teeth are in a single series, and this is the chief reason for placing these species in distinct genera. The dorsal fin has nine, exceptionally eight or ten, branched rays, the anal twenty-three to twenty-nine; the lower caudal lobe is notably longer than the upper. There are forty-nine to fifty-seven scales in the lateral line, eleven to fifteen in a transverse series from the origin of the dorsal fin to the lateral line, and six to seven and a half between the lateral line and the base of the pelvic fin. Young examples are similar in colour to the White Bream, and are generally confounded with that species under the name "Bream-flat"; the adults are greenish or brownish, with brassy reflections on the sides and with blackish fins.

The Bream is found in Europe north of the Pyrenees and the Alps, in Russian Turkestan, and in Western Siberia. In our islands it is absent from Scotland north of Loch Lomond and the Firth of Forth, the western parts of Wales, and

PLATE XXIX

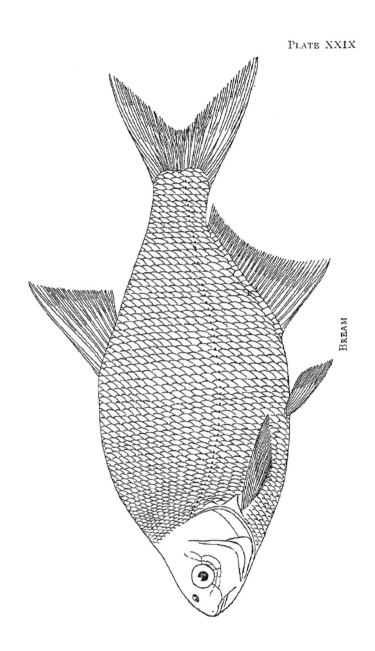

from Dorset, Somerset, Devon, and Cornwall; it is common in Ireland.

Bream of 8 or 9 lbs. are considered large, but there is a record of a 17-lb. fish from the Trent, and a still greater size is attained on the Continent. One a little less than a foot in length is figured on Pl. XXIX.

Like the White Bream, this species is found only in lakes or sluggish streams; it is especially abundant in the Norfolk Broads and in some of the Irish lakes. It swims in large shoals, and on warm summer days a number of these fish may sometimes be seen lying motionless near the surface of the water, as a rule, however, they keep on or near a weedy or muddy bottom in fairly deep water; here they feed greedily on the weeds or on organic matter which they extract from the mud, also on insect larvæ, little shellfish, etc.

The Bream is of a shy and cunning disposition; it is said sometimes to escape its enemies by stirring up the ooze in order to cloud the water, and to lie flat on the bottom or burrow in the mud so that the seine passes over it. The spawning usually takes place in May, when the shoals repair to shallows near the banks, making a great noise, swimming near the surface, leaping, splashing, and rolling about. The females deposit their roe on the weeds and rushes, against which they rub themselves; after a few days the whole shoal make their way back into the deep water. The eggs are very numerous; according to different authors they hatch in from one to three weeks; the young Bream are often found on the shallows in company with the White Bream, from which the fishermen usually fail to distinguish them; they are said to attain sexual maturity when they are four years old and about a foot long.

The Bream is not much esteemed as food, although it has a certain value from its numbers, the size it attains, and the ease with which it can be kept in ponds or transported alive; nowadays it is more appreciated on the Continent than in this country.

The word Bream or Breme is from the old French name *Brême*, the derivation of which is uncertain.

That comparatively closely related species, such as the Roach and Rudd, should form a natural hybrid is not surprising, but that the fishes of the Bream group, characterized by the long anal fin and by the compression of the abdomen behind the pelvic fins to an edge which divides the scales, should hybridize with the very different Roach and its allies is distinctly peculiar.

The Bream and Roach hybrid is known from nearly all parts of the area inhabited by the parent species; it is represented in the British Museum by a number of specimens, including some from Tetworth, the Nen at Lilford, Norwich, the canal at Slough, the Colne, and the Avon. The largest of these, from Lilford, is more than a foot long; a little specimen from Norwich was sent in 1903, with some young Bream, as representatives of the "Bream-flat" of Norfolk. Very remarkable was the experience of Messrs. S. Ling and J. Ladbroke, who in one afternoon's fishing in the canal at Slough captured seven examples of this hybrid, all 7 to 8 inches long, and nothing else. A specimen 10 inches long is shown on Pl. XXX, Fig. 1.

PLATE XXX

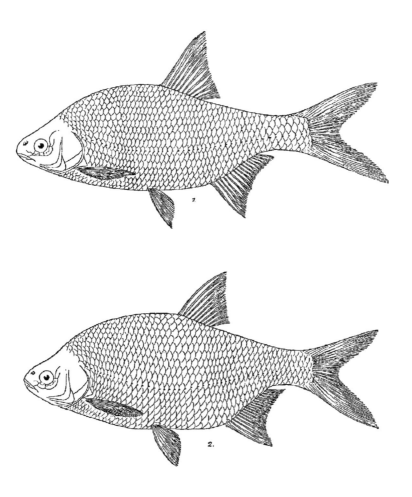

1. HYBRID ROACH AND BREAM; 2. HYBRID RUDD AND BREAM

In form this fish resembles a rather deep Roach, the greatest depth of the body measuring one-third to two-fifths of the length to the base of the caudal fin. The snout is blunt and the mouth is nearly terminal, with horizontal or slightly oblique cleft and with the lower jaw a little shorter than the upper. The dorsal fin is situated above the interspace between the pelvic and anal fins, and the anal is rather elongate, with fifteen to nineteen branched rays. The abdominal keel between the pelvic and anal fins is sometimes weak and covered with angularly bent scales, sometimes sharper, with a naked line separating the scales of the two sides. The scales number forty-seven to fifty-two in the lateral line, nine to eleven in a transverse series from the origin of the dorsal fin to the lateral line, five or six between the lateral line and the base of the pelvic fin.

The numerical characters of the parent species and of the hybrid may be tabulated as follows :—

	Bream	Hybrid	Roach
Pharyngeal teeth . . .	5–5	5 or 6–5 or 6	5 or 6–5 or 6
Branched rays in dorsal fin .	8–10	9–11	9–11
,, ,, anal fin .	23–29	15–19	9–12
Scales in lateral line . .	49–57	47–52	40–46
,, from origin of dorsal fin to lateral line .	11–15	9–11	7–9
,, from lateral line to base of pelvic fin .	6–7	5–6	3–4½
Vertebræ . . .	43–45	42	39–41

The hybrid between the White Bream and the Roach is known from Holland, Belgium, and Germany, but not from England. One from

Bavaria in the British Museum has eight branched rays in the dorsal fin and thirteen in the anal, forty-six scales in the lateral line, nine and a half between the origin of the dorsal fin and the lateral line, and four and a half between the lateral line and the base of the pelvic fin. The pharyngeal teeth are in two series, 1.5—5.1.

The hybrid between the Bream and the Rudd was little known until I described it a year or two ago from fourteen examples in the collection of the British Museum; nine were sent to me from Lough Erne by Major Trevelyan, and one of these, nearly 10 inches long, is figured (Pl. XXX, Fig. 2); three from Colebrooke, Upper Lough Erne, were received in 1871 from Sir Victor Brooke; one from Thetford was presented by Dr. Gunther in 1879, and one is a skin from Yarrell's Collection. The Bream × Rudd hybrid differs from the Bream × Roach hybrid in the same way that the Rudd does from the Roach. The body averages deeper, the greatest depth being contained two and one-fourth to two and two-third times in the length to the base of the caudal fin. The mouth is more oblique, with the jaws equal or the lower the shorter. The dorsal fin is farther back and usually has fewer rays, nine (eight to ten) instead of ten (nine to eleven) branched ones. The pharyngeal teeth are usually in two series, and the main row never has more than five teeth.

The numerical characters of this hybrid and of the parent species may be contrasted with those of the Bream × Roach hybrid and its parent forms given above :—

	Bream	Hybrid	Rudd
Pharyngeal teeth . . .	5–5	5–5 to 2.5–5.2	3.5–5.3
Branched rays in dorsal fin .	8–10	8–10	8–10
,, ,, anal fin . .	23–29	15–18	10–13
Scales in the lateral line .	49–57	46–50	39–44
,, between origin of dorsal fin and lateral line .	11–15	9½–10½	7–8
,, between lateral line and base of pelvic fin. .	6–7	3½–5½	3–4
Vertebræ.	43–45	42	37–39

According to Major Trevelyan, this fish is known to the Lough Erne fishermen by the name of " White Roach," in contradistinction to the " Red Roach " or " Rudd." The largest specimen sent by him measured 13 inches in length and weighed 2 lbs., but he had good reason to believe that specimens of 2½ lbs. had been taken. The abundance of these fishes in Lough Erne is very remarkable, and it would be interesting if it could be ascertained whether they attempt to breed, and if so, with what effect. Evidently in Lough Erne the breeding seasons of the Bream and Rudd usually coincide.

Yarrell's skin and the Colebrooke and Thetford examples had been determined as White Bream (*Blicca bjœrnka*), a species in many ways intermediate between the Rudd and the Bream, but differing from their hybrid in the less oblique mouth, larger eye, different number of fin-rays (dorsal usually with eight instead of nine branched rays, anal with nineteen to twenty-four), and higher position of the lateral line, which runs at about two-fifths instead of one-third the height of the body in the middle of the length of the fish.

16

Here may be mentioned the hybrid White Bream and Rudd, known on the Continent and perhaps to be met with in our eastern counties This differs from the Bream × Rudd hybrid in the same way that the White Bream does from the Bream, *i.e.* the smaller size (maximum length 10 inches), the larger eye, the fewer scales (forty to forty-six in the lateral line, eight or nine from origin of dorsal fin to lateral line), fewer fin-rays (dorsal usually with eight, anal with twelve to seventeen branched rays), and more numerous pharyngeal teeth (2.5–5.2 to 3.6–5.3).

THE BLEAK (*Alburnus lucidus*) is closely related to the Breams, differing especially in the elongate form of the body and the strongly oblique mouth with the lower jaw projecting.   In this species there are seven to nine branched rays in the dorsal fin, and fifteen to twenty in the anal; the scales in the lateral line number forty - six to fifty - four.   The coloration is silvery white, with the back greenish.

The Bleak is found all over Europe north of the Pyrenees and the Alps ; it is absent from Scotland and Ireland, but is widely distributed in England and Wales, although it does not occur in the Lake District, West Wales, or in the counties bordering the English Channel.

This is a small species, and specimens of 8 inches are rare ; that figured on Pl. XXXI, Fig. 1, measures nearly 6 inches.   It is a pretty and lively little fish, living in shoals and preferring clear running water ; throughout the summer it sports at the surface, darting about and leaping at the flies, and feeding also on any small worms, shrimps, or larvæ which it may find.

PLATE XXX

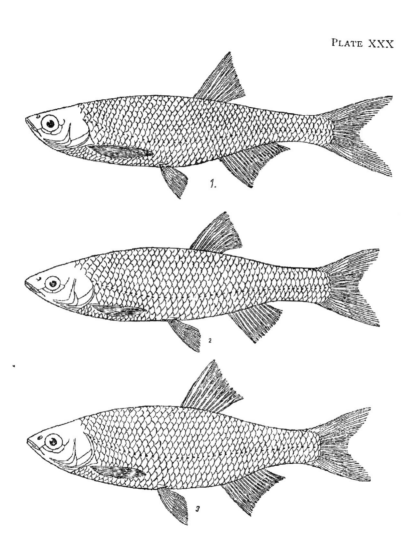

1. BLEAK; 2. HYBRID CHUB AND BLEAK; 3. HYBRID ROACH AND BLEAK

The breeding season is usually in June, but varies according to the year and the locality; and also according to the age of the fish, the older ones spawning earliest. At this time the Bleak press in close to the bank in vast shoals, leaping and lashing the water with their tails, thus producing a sort of hissing noise; in their excitement some may even jump ashore and so perish. The eggs are shed in the shallow water and adhere to the weeds or the stones on the bottom.

The small size of this fish makes it scarcely worth the attention of the angler, especially if he be in search of something to eat, however, it is an excellent spinning bait for Pike, which are attracted by its bright coloration. The silvery scales of the Bleak are extensively used in the manufacture of artificial pearls, especially in France; hollow glass beads are coated on the inside with the pigment obtained by scraping the scales, and are then filled in with wax. This industry commenced in France in the year 1656, but the Chinese are said to have extracted the silvery crystals from the scales of fish for centuries before this.

The Bleak takes its name fiom its brilliant white colour; the word bleach seems to be from the same root.

Forms interpreted as hybrids between the Bleak and the White Bream, Rudd, Roach, Dace, and Chub have been described by various authors, but only two of these are known from England.

The Roach × Bleak hybrid is known only from a specimen $5\frac{1}{2}$ inches long, taken in the Nene, which is here figured (Pl XXXI, Fig. 3). It has a moderately deep body, the depth being contained

three and a half times in the length to the base
of the caudal fin. The mouth is terminal and
oblique, with the jaws equal anteriorly. The dorsal
fin, of three simple and nine branched rays, is
above the interspace between the pelvic and anal
fins, and the branched rays in the last-named
number thirteen; the free edge of the dorsal and
anal fins is concave. The abdomen between the
pelvic fins and the vent is compressed to a ridge,
which is crossed by the scales except posteriorly.
The Roach parentage is conclusively established
by the pharyngeal teeth, which as in that species
are hooked, slightly denticulated, set in a single
series, five on one side and six on the other. There
are forty-three scales in the lateral line, eight in
a transverse series above it, and four between it
and the base of the pelvic fins.

The Chub x Bleak hybrid has been recorded from
various continental localities, and is also represented
in the British Museum collection by specimens from
the Thames at Staines, the Mole at Moulsey, and
a reservoir near Oundle. In these fishes the body
is elongate, as in the Chub and Bleak; the mouth
is terminal and oblique, with the jaws equal anteriorly; the dorsal fin, composed of three simple
and eight branched rays, is situated above the
interspace between the pelvic and anal fins, whilst
the branched rays in the anal fin number ten to
thirteen, more than in the Chub, which has eight
or nine, and fewer than in the Bleak, which has
from fifteen to twenty. The belly behind the
pelvic fins is compressed to a sharp edge, but the
scales pass over this except posteriorly.

The features so far described show that the

Bleak is one of the parents, but leave it in doubt whether the Chub or the Dace is the other; the Chub differs from the Dace in the larger mouth and broader head, whilst the scales are somewhat fewer and the free edges of the dorsal and anal fins are convex instead of concave. All these characters come out to a greater or less extent in the hybrid fishes here described and establish their Chub parentage pretty conclusively. The mouth is a little larger than in either the Dace or the Bleak, the length of the lower jaw measuring more than one-third that of the head, the inter-orbital space is broader than in the Dace and much broader than in the Bleak, its width being contained two and three-fourth to two and four-fifth times in the length of the head; the scales in the lateral line number forty-five to forty-eight, averaging forty-six, whereas the average number in Dace and Bleak is about fifty; the free edges of the dorsal and anal fins may be either very slightly concave, straight, or a little convex.

This hybrid sometimes attains a length of a foot, a size which also points to the Chub rather than the Dace as one of the parents; the example figured (Pl XXXI, Fig. 2) is nearly half that length.

THE LOACHES (*Cobitidæ*) are closely related to the Carp family, from which they differ in having the reduced air-bladder partly or wholly enclosed in a bony capsule, whilst the pharyngeal teeth, arranged in a single series, are rather more numerous than in the Cyprinoids, and the pad against which they work in the latter group is absent. Externally the Loaches may be recognized by the elongate body, with the scales very small or absent, and by the presence of at least six barbels.

The fishes of this family are small, and numerous species occur in the mountain streams of Central and Southern Asia; one species is known from Abyssinia and three inhabit Europe, two of which are found in our islands.

It is well known that Loaches use their intestine as an accessary organ of respiration; when the water is low or stagnant they come to the surface and swallow air, which they afterwards expel through the vent.

Loaches are very sensitive to changes of atmospheric pressure, so much so that in Germany one species (*Misgurnus fossilis*) is called 'Wetterfisch,' foretelling stormy weather by becoming restless and frequently rising to the surface of the water. This peculiar sensitiveness is doubtless due to the structure of the air-bladder, enclosed in a bony capsule with an orifice on each side, from which a duct, filled with a gelatinous substance, passes between the main masses of the trunk muscles and ends immediately beneath the skin just above the base of the pectoral fin. Thus the internal ear is brought into communication with the exterior *via* the Weberian ossicles and the air-bladder.

In the three European species of Loach, and probably in many others, the air-bladder has no open duct, and like many Cat-fishes with a reduced and encapsuled air-bladder, they form an exception to the statement that the fishes of the order Ostariophysi are phyostomous.

THE LOACH or STONE LOACH (*Nemachilus barba-*

PLATE XXXII

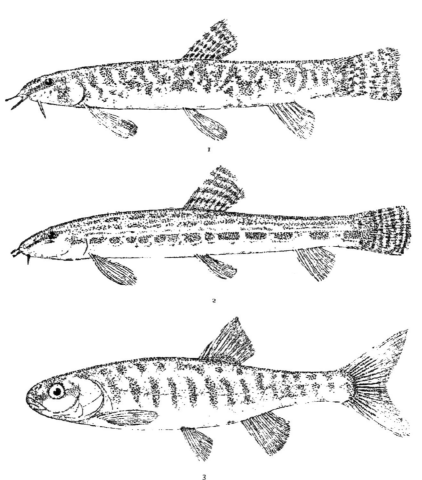

1. LOACH; 2. SPINED LOACH; 3. MINNOW

tula) (Pl. XXXII, Fig. 1) has the body elongate and
subcylindrical, and the head rather long, with the
small eyes placed high up, and the blunt snout
projecting beyond the transverse inferior mouth, in
front of which are two pairs of barbels, whilst a
third pair are placed near its corners. The gill-
openings are small, ending below in front of the
base of the pectoral fins; the pharyngeal bones
bear a series of ten or fewer small pointed teeth
on each side. The dorsal fin, of three simple and
seven or eight branched rays, is placed in the middle
of the length of the fish, and its origin is above
or a little in advance of the base of the pelvics;
the anal has three simple and five or six branched
rays, and the caudal is subtruncate. The scales
are so minute and hidden by slime that the body
generally appears naked; the lateral line may be
seen on the middle of the side. The colour of
the back and sides is usually greyish, sometimes
yellowish, brownish, or greenish, with darker spots
and marblings; the belly is whitish, and the dorsal,
caudal, and pectoral fins are barred with series of
small blackish spots.

The usual length of this little fish is 3 or 4
inches, and 5 inches is about the maximum. It
is widely distributed, being found all over Europe,
except the Iberian Peninsula and Greece, and in
Asia extending through Turkestan and Siberia to
northern China, Corea, and Japan. In our islands
it is generally present, except in the Northern
Highlands of Scotland.

As a rule the Loach is found in small clear
streams with a sandy or gravelly bottom, and
spends most of its time during the day concealed

beneath a stone, from which it darts off rapidly when disturbed to the nearest refuge at hand. It is said to be more active at night, and feeds chiefly on worms, shrimps, insect larvæ, etc.

The spawning season is usually in April or May; the only account of the breeding habits of this species with which I am acquainted is that of Knauthe, which is sufficiently interesting to be translated. He writes· "As I was strolling along the banks of a clear and rapid mountain brook in April, 1887, I noticed some Loaches in a deep hole which had been hollowed out by the water; they kept swimming round near the edge, the males, which were usually the smaller and weaker, always following the stronger females. When they had circled the pool several times, all of them, in number perhaps about a dozen, swam to the projecting roots of an old willow. Here a female fish forced her way through a hole, formed by three roots, and about equal in diameter to the thickness of a finger; one or more males immediately followed her. The other Loaches occupied themselves in the same way, using similar openings, of which there were plenty beneath the surface of the water. Owing to the considerable irritation of the belly, the females shed their eggs and the males their milt."

The Loach is said to be excellent eating, but in this country is not now much sought after for this purpose, and its chief use is as a bait for Perch and Pike.

THE SPINED LOACH (*Cobitis tænia*) (Pl. XXXII, Fig. 2) is especially remarkable for the mobility of the præfrontal bone, which has the form of an erectile bifid spine; when at rest this lies in a groove below the eye. The body is more compressed than in the Stone Loach, and the coloration is different, there being a regular series of dark brown or blackish oblong or oval spots along each side of the body.

This is generally a smaller species than the preceding, but may attain a length of 4 inches or more. It is found all over Europe, and also extends through Siberia to Corea, northern China, and Japan. In the British Isles it may perhaps not be so rare as is generally supposed, for on account of its habits it is not easily observed; it has been recorded from a few localities in England, but is not known from Scotland, Wales, or Ireland.

It prefers clear brooks with a sandy or gravelly bottom, and often lies still among the stones, or buried in the sand with only the head exposed. Like the Stone Loach it is more active at night, and feeds on worms, shrimps, and larvæ. It uses the pair of spines on the head as weapons, and if caught in the hand will erect them and stab into the flesh.

This species spawns at about the same season as the Stone Loach; it is said to be worthless as food, and has no value for any other purposes, so that its scarcity in our waters need not concern us greatly.

CHAPTER XI

THE BURBOT, THE PERCHES, AND THE GREY MULLETS

The Anacanthini. The Burbot described—distribution—size —food and habits—breeding—resemblances to Eel and Bull-head —value as food—names The Percomorphi. The Perch family . the Perch described — distribution —size — malformations — variation in colour—food and habits — breeding — growth— tenacity of life — extraordinary boldness — origin of name. The Ruffe . differences from Perch—distribution—size—habits —names Perch × Ruffe hybrid. The Bass differences from Perch — size — distribution — habits. Black Bass. The Grey Mullets the Thick-lipped Grey Mullet—the Golden Grey Mullet—the Thin-lipped Grey Mullet—migrations, food, and habits of Grey Mullets—their boldness and cunning—breeding— growth of young—value as food

THE BURBOT or EEL-POUT (*Lota lota*) is the only freshwater fish of the Cod family (*Gadidæ*) or of the order Anacanthini, a group which includes fishes with all the fin-rays flexible and jointed, but which differ rather widely in other respects from the soft-rayed fishes previously considered. In all the Anacanthini the air-bladder has no open duct, the mouth is toothed and protractile, with the præmaxillaries excluding the maxillaries from the gape, and the pelvic fins, often many-rayed, are placed below or in advance of the pectorals, although the pelvic bones are not directly attached to the clavicles. In this order the true caudal fin, corresponding to the caudal fin of other fishes, is either absent or else greatly reduced and united with the dorsal and anal; the so-called caudal fin of the Burbot and other members of the Cod family is mainly formed of dorsal and anal rays.

The Burbot has the body elongate, subcylindrical anteriorly and compressed posteriorly, with the abdomen rather prominent; the head is broad and depressed, with the eyes placed well forward and far apart. The mouth is wide, with bands of small pointed teeth in the jaws and a crescentic band of similar teeth on the vomer; a barbel depends from the middle of the lower jaw, whilst the edge of each anterior nostril is produced into a small barbel. The dorsal fin is composed of numerous rays and is divided into a short anterior and a long posterior part, the latter opposite to the anal and, like it, continuous with the rounded terminal fin. The pelvic fins are six-rayed, are widely separated, and are placed in advance of the pectorals. The scales are very small and the thick skin is covered with a slimy mucous secretion. The coloration is yellowish, greyish, brownish, or greenish, spotted or marbled with dark brown or black on the back and sides, and shading below into white or pale yellow.

The Burbot is found in the fresh waters of Europe, except the Iberian Peninsula and Greece; it ranges throughout Siberia, and in North America occurs in the region of the Great Lakes and northwards. In Britain it appears to be confined to rivers flowing into the North Sea, from Durham to Norfolk.

As a rule in our waters it does not attain a length of more than 2 feet, with a weight of about 3 lbs.; a specimen of 8 lbs. from the Trent seems to be the largest recorded English Burbot, but double this size is attained on the Continent, and in the arctic regions they grow very large, Burbot weighing as much as 60 lbs. having been taken in Alaska. The small example figured (Pl. XX, Fig. 1), 10 inches long, is from the Trent.

This is an inhabitant of clear rivers and lakes, living at the bottom in deepish water; here it lies, usually more or less concealed among the weeds or in crevices between stones, or sometimes in holes under the banks among the roots of trees. It is a gluttonous fish, and eats great quantities of the eggs of other species; it devours greedily almost anything edible that can be obtained, but is especially destructive to other fish; in the day-time the Burbot pounces on any little fish un-fortunate enough to come near enough to its lurking-place, whilst at night it goes in active pursuit of prey. Even fish of some size may fall victims to its voracity, and a case has been recorded of a Burbot of 22½ inches which had swallowed a Pike of more than half its own length.

The breeding season is usually from January to March, when the Burbots assemble in shoals. Some large fish may spawn in deep water, but the majority repair to the shallows; the eggs are very small and numerous, and are deposited on the bottom. They hatch out in three or four weeks, and the fry grow but slowly, attaining a length of about 4 inches in a year; young Burbots may often be found under stones on the shallows;

when three or four years old they become sexually mature.

In its mode of life the Burbot recalls the Eel and the Bull-head, and whilst its head and mouth remind us of the latter, its long slippery body and its undulating movements when swimming are points of resemblance to the former species. It is readily captured in traps or on lines baited with worms or small fish, being neither shy nor wary; it is much appreciated as food, especially on the Continent, the flesh being white and firm, and the liver in particular being esteemed a delicacy. It is very tenacious of life and will live for some time out of the water.

The name Burbot is from the old French *Bourbotte* (modern French *Barbote*), whilst Eel-Pout is from the Saxon *Aele-puta*, a name alluding to the Eel-like form of the fish and probably to its prominent, pouting belly.

We now come to the important order PERCO-MORPHI, including fishes which agree with the Anacanthini in the absence of a pneumatic duct and in the structure of the mouth, which is typically toothed, with the protractile præmaxillaries excluding the maxillaries from the oral border. The Percomorphi, however, differ from all the fishes previously considered in fin structure; the anterior rays of the dorsal fin are unjointed, typically stiff and pungent spines, which often form a distinct spinous dorsal fin; the anal fin is preceded by similar spines; the pelvic fins are placed anteriorly, and each is formed of six or fewer rays, the outer of which is usually spinous. As a rule the pelvic bones are directly attached to the clavicles, but in some members of the order this connexion has not been acquired, or has been lost.

The fishes of this order which occur in the fresh waters of our islands may be primarily divided into those with normal suborbitals and those with the second suborbital produced across the cheek to the præ-operculum. The former includes the Percoids, or Perch-like fishes, and the Mugiloids, or Grey Mullets and their allies. To the latter group belong the Scorpænoids and the Gastrosteoids, which are dealt with in the next chapter.

The Percoids, or Perch-like fishes, represent the highest stage in the evolution of the normal piscine type, although various aberrant offshoots of the group have become adapted to peculiar modes of life.

The Perch family (*Percidæ*) includes a number of fishes inhabiting the lakes and rivers of North America, Europe, and northern Asia. The majority of the species are little fishes found in the rivers of the United States and known as " Darters," whilst the largest members of the family are the Pike-perches (*Lucioperca*), some species of which attain a length of 4 feet, predaceous fishes closely allied to the Perch, but more slender, and with a larger mouth and stronger teeth.

THE PERCH (*Perca fluviatilis*) has the body somewhat compressed, and moderately elongate, tapering posteriorly, with the back more or less humped between the occiput and the dorsal fin; the upper part of the head is smooth and scaleless, whilst the body is covered with small, rough, adherent scales. The mouth is rather large, with pointed teeth in

PLATE XXXIII

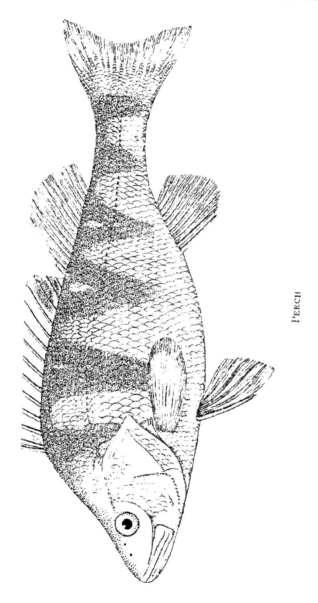

PERCH

the jaws and on the palate, the opercle ends in a sharp spine. There are two dorsal fins, a longer anterior one formed of from fourteen to sixteen spines, and a shorter posterior one with two spines and thirteen to fifteen soft rays; the anal has two spines and eight to ten branched rays; the caudal is emarginate; each pelvic fin consists of a strong spine and five branched rays; they are inserted close together a little behind the vertical through the base of the pectoral.

The colour of the back and sides is greenish olive, shading below into a golden yellow, whilst the belly is white; on the sides appear five or more dark vertical bars, which are sometimes bifurcated above, the spinous dorsal is greyish, with a black spot behind, and the lower fins are reddish. Sometimes the bars are inconspicuous and the whole fish appears almost uniformly dark greenish, sometimes the deep black bars are in striking contrast to the pale olive ground colour.

The Perch is found all over Europe, except the Iberian Peninsula, and in Transcaspia and throughout Siberia. Two closely allied species represent it in eastern North America and in Turkestan respectively. In our islands it is generally distributed, except in the Northern Highlands of Scotland.

It does not usually attain a length of more than 18 inches, with a weight of 4 to 5 lbs., but much larger specimens have occasionally been met with. I have never caught one of these monsters, but my angling experiences lead me to believe that there is no reason to doubt the authenticity of records of Perch of 6 and even 8 and 10 lbs., captured in

17

our waters; one of the last weight is said to have been caught in Bala Lake, and one of 8 lbs. was taken from the Avon in Wiltshire by means of a night-line baited with a Roach. Our figure (Pl. XXXIII) is of a 1-lb. fish from Lough Erne.

In some localities abnormal Perch are not uncommon; a hunch-backed form with a very deep body, resembling a Crucian Carp in shape, and paralleling the Hog-backed Trout, occurs in Loch Arthur in Kirkcudbrightshire, Llyn Rhaithlyn in Merionethshire, etc.; it has also been taken in Cheshire and in various Scandinavian lakes; this hunch-backed appearance is the outward manifestation of a malformed vertebral column, which is shortened and has the number of joints reduced by the imperfect development and fusion together of several of the vertebræ. In various places Perch may be taken with the snout very short and blunt and the lower jaw projecting; similarly deformed Salmon, Trout, and Pike have also been captured in our islands. A Perch from Coggeshall, Essex, in the British Museum collection, shows a reverse malformation, the snout projecting beyond the lower jaw, which is very short.

The coloration of the Perch varies somewhat according to locality, age, sex, and season; the young are usually paler than the adults and the males more brilliant than the females, whilst all are darker in the winter than in the summer; in the latter season it is a strikingly handsome fish, and a large Perch about to seize its prey, with fins spread and mouth agape, forms a fine picture.

Perch live in shoals, in rivers, lakes, and ponds; they prefer deep pools or places where the current is slow, and are fond of swimming near beds of reeds, or where there are sunken trees or piles driven into the water, places where the fry of other fish are to be found. In Norfolk they rather like a little salt water, and the largest are said to be taken where the water begins to turn brackish. They are bold and voracious, and in the summer months rove in company over the shallows in active pursuit of their prey, which consists of worms, insect larvæ, shellfish, etc., and of little fish such as Minnows, Gudgeon, and Bleak, and the fry of other species; it is especially in the early morning and in the evening that they go in search of food, whilst in the middle of a warm day a whole troop may often be seen lying motionless in mid-water, or sometimes quite near the surface; in the winter, however, they keep to the deep water and feed only in the middle of the day. If it lacks caution the Perch makes up for it in courage, and when of any size is well able to take care of itself, erecting its dorsal spines and facing the foe which threatens an assault.

In the spring the Perch forsake the deeps wherein they have passed the winter, and migrate in shoals to sandy or gravelly shallows, selecting places where reeds or rushes grow in the water or where there are sunken branches. Here they breed, the season varying from March to May according to the locality and the weather. The roe is held together by a membrane, and the females rub against a stone or twig or amongst the reeds, until the end of the band of eggs has become attached, and then dart forward, twisting from side to side so that the roe comes out in a string, when it is fertilized by the attendant males. The bands of eggs, attached at one end, float in the water, enveloped in a coat of mucus; in a few days the fry appear, and after resting at the bottom for about a month, during which the small yolk-sac is absorbed, they swim about at the surface and feed on minute organisms; after a year's growth they are 3 or 4 inches long, whilst two-year-old Perch usually measure 5 or 6 and those of three years 7 or 8 inches; the last are sexually mature.

As a rule a shoal of Perch is composed of fish of nearly the same size and age, and it usually happens that a much greater number of small Perch swim in company than is the case with the larger ones. The Rev. R. Lubbock recorded the capture of eighteen Perch in one place on the Broads, not one of which weighed less than 2 lbs.; this illustrates well the uniform size of a shoal.

Perch are very tenacious of life, and can be sent alive for long distances packed in damp grass, in continental markets they are often exhibited, and if not sold are put back in the ponds from which they were taken in the morning.

The Perch is an excellent food-fish, with white, firm, and well-flavoured flesh; those taken in clear rivers are better for the table than pond-fish. It affords good sport to the angler, but is overbold, and except when large does not require much skill to be expended on its capture; small ones are sometimes used as bait for Pike.

As an example of the boldness of this fish, Mr. Pennell instances the case of a small Perch which he captured, and in removing the hook accidentally displaced one eye; he returned the fish to the lake and threw in the line with the eye remaining on the hook and with no other bait; he immediately recaptured the fish he had just thrown in, which had actually been caught with its own eye.

The name is derived from the Latin *Perca* through the French *Perche*; it is of Greek origin and appears to mean either "dark coloured" or "spotted." Barse or its modern equivalent Bass is from the Anglo-Saxon *Baers*, the older name for this fish in our islands, which is still used for it locally, for example in Westmorland and Norfolk.

THE RUFFE or POPE (*Acerina cernua*) differs from the Perch in the confluence of the dorsal fins to form a single fin with thirteen to sixteen spines and eleven to fifteen soft rays, the shorter anal fin, of two spines and five or six soft rays, the rather wide interspace between the bases of the pelvic fins, the larger scales, the small size of the mouth, the presence of large muciferous cavities on the head, the fewer and stronger præ-opercular spines, and the colour, which is greenish or yellowish olive, marbled and spotted with brown or black

This species inhabits Europe north of the Pyrenees and the Alps, Russian Turkestan, and Siberia; in our islands it is absent from Scotland and Ireland; in England and Wales it extends north to Lancashire and Yorkshire, and west to the Severn and Dee; it is absent from Somerset, Hampshire, Dorset, Devon, and Cornwall. In our waters it does not exceed a length of 7 or 8 inches, and one is shown of the natural size on Pl. XXXV.

The Ruffe is found in lakes, canals, and slow-running rivers, usually swimming in shoals and keeping near the bottom in fairly deep water; when resting it is not easily distinguished, even when the water is clear, so closely does its speckled coloration resemble the sand or gravel. It is a sluggish fish, feeding on small fry, worms, insects, shellfish, etc., but rarely troubling to go in active pursuit of food. It is especially plentiful on the Broads and in the rivers and canals of our midland and eastern counties; the Rev. R. Lubbock wrote that in Norwich hundreds might be seen on a summer's day, at the piers of the bridges, hunting for worms and slugs amongst the weeds on the piles, introducing the head and half the body until a part of the pier appeared studded with their tails.

In March or April the shoals of Ruffe migrate from the deep and still waters wherein they have passed the winter to the streams, or to fairly shallow places near the banks of the lakes, especially choosing spots where reeds or sedges grow in the water; here they spawn after the same fashion as the Perch.

Like the latter fish, the Ruffe is very tenacious of life, and it is held in even better repute as food, the flesh being white, firm, of good flavour, and easily digested; its small size renders this fish of little account to the angler, to whom it is sometimes rather a nuisance, as it bites boldly, like its larger relative the Perch.

The name *Ruffe* is an alternative spelling of 'rough,' and was no doubt given to the fish from its roughness, due to the presence of little denticles near the free edges of the adherent scales. *Pope* is of unknown origin and is said to have been applied as a term of contempt; the cruel custom of pressing a cork on to the dorsal spines and returning the fish to the water, thus condemning it to a lingering death, as it must swim at the surface and cannot find food, may possibly have originated in hatred of papistry.

It is of interest to note that two such very distinct species as the Perch and the Ruffe form a natural hybrid,[1] which has been obtained in the Danube, although it has not yet been recorded from our waters. In the number of scales and fin-rays this hybrid is intermediate between the parent forms, but experimental breeding has shown that whilst the offspring of a Ruffe father and Perch mother are intermediate in other characters also, as in form and coloration, showing the speckling of the Ruffe and the bars of the Perch, yet the progeny of a male Perch and a female Ruffe resemble the mother in form and coloration, and are not very easy to distinguish from pure-bred Ruffes. These hybrids are not fertile *inter se*, but are quite fertile with either parent.

THE BASS (*Morone labrax*) belongs to the Sea-perch family (*Serranidæ*), which differs from the Perch family in that the anal fin has three spines instead of one or two, whilst the eye is supported by an internal laminar process of the suborbital bones.

[1] Kammerer, *Arch. Entwickel. mech. Leipzig*, xxiii., 1907, p. 511.

The six species of the Bass genus inhabit the North Atlantic and Mediterranean and the rivers which flow into them, and it is interesting to note that whilst the two European species are marine, only ascending rivers occasionally, two of the North American forms are anadromous, spawning in fresh water, whilst the remaining two have become permanent freshwater residents.

The Bass shows many structural resemblances to the Perch, from which it is at once distinguished by the coloration, which is nearer to that of the

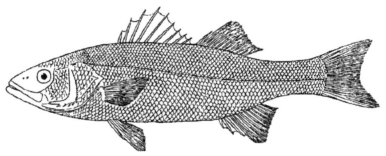

FIG. 21.—Bass.

Salmon, being silvery, with the back bluish grey or olive; a dark spot is present on the opercle, and small scattered blackish spots may be present in the young. The Bass also differs notably from the Perch in the stronger and fewer spines, eight to ten in number, of the spinous dorsal fin, and in having three anal spines instead of two.

This species grows to a much larger size than the Perch, attaining a length of more than 3 feet and a weight of nearly 30 lbs., although fish of half that size are considered large. It is found in the Mediterranean and on the Atlantic coasts of Europe, becoming rare northwards. In the British Isles it is most abundant on the southern coasts, and in the summer months the shoals may journey up the rivers for considerable distances, in Sussex, for example, ascending the Arun to Pulborough and the Cuckmere River to Alfriston.

The Bass is a voracious fish, and the shoals go in active pursuit of their prey, which is chiefly small fishes and crustaceans. It has long been celebrated for its cunning and is difficult to capture in nets, seizing any opportunity of escape which offers, swimming under the foot-rope if the bottom be uneven, or burrowing when the bottom is level and sandy. This species spawns in the sea near the coasts, usually rather late in the summer, and the eggs are pelagic.

Bass afford fine sport and also much vexation to the angler, for in addition to their strength and cunning they are very capricious, and it is sometimes almost impossible to hit on the right bait. As food they have been held in much esteem since the days of the Romans, who kept them in freshwater aquaria, a practice which is said to improve the flavour.

The word Bass is a corruption of the middle English *Barse*, from *Baers*, the Anglo-Saxon name for the Perch.

The Black Bass (*Micropterus*) belong to an allied family (*Centrarchidæ*), which is confined to the rivers of North America. There are two species, the Large-mouthed and Small-mouthed Black Bass, which have been introduced with more or less success into some of our English rivers.

THE GREY MULLETS (*Mugilidæ*) belong to the group of Mugiloids, which differs from the Percoids in a single character—the insertion of the pelvic fins well behind the base of the pectorals so that the pelvic bones are not attached to the clavicles; whether they are on this account to be regarded as more primitive or more specialized than the Percoids is not quite certain.

The Sand-smelts (*Atherinidæ*) are closely related to the Grey Mullets, but the two other families of this group are very different, the Barracudas (*Sphyrænidæ*) being large carnivorous Pike-like fishes, almost the most dangerous inhabitants of the tropical seas, whilst the Thread-fins (*Polynemidæ*) are peculiar in that the lower pectoral rays are detached from the rest of the fin and are produced into long filamentous feelers.

Numerous species of Grey Mullet are known from temperate and tropical seas, frequenting bays and estuaries and often entering fresh water. Our British species are three in number, in all the body is moderately elongate, slightly compressed, the head short and broad, the mouth small, terminal, protractile, bordered above by the præmaxillaries only; the setiform teeth are so feeble that they appear merely as a fringe on the jaws; the eyes are lateral, placed high, but more visible from below than from above. The head is scaly and the scales on the body are large, numbering forty to fifty in a longitudinal, and about fourteen in a transverse, series; there is no lateral line. The first dorsal fin is formed of four pungent spines, the second of a spine and eight or nine soft rays; the anal originates somewhat in advance of the dorsal and has three spines and from eight to ten soft rays; the pelvics are inserted well behind the base of the pectorals, which are placed high. The colour is silvery grey, with the back bluish and the belly white; on the sides there are more or less pronounced dark longitudinal stripes along the series of scales.

THE THICK-LIPPED GREY MULLET (*Mugil chelo*) (Pl. XXXIV, Fig. *a*) has the upper lip thick and more or less distinctly papillose, especially in the adult fish; when the mouth is closed, the distal ends of the maxillaries are exposed, whilst the rami of the lower jaw are separated by a narrow inter-space, or by none; the pectoral fin measures at least three-fourths the length of the head. This is the commonest British species; it ranges from Scandinavia to the Mediterranean; it has been known to attain a length of 3 feet and a weight of nearly 15 lbs.

THE GOLDEN GREY MULLET (*Mugil auratus*) (Pl. XXXIV) differs in that the upper lip is not so thick; when the mouth is closed the maxillaries are almost or quite concealed, and the rami of the lower jaw are separated by an elliptical area. This fish takes its name from a pair of golden spots on each side of the head, one on the operculum and a smaller one in front of it, behind the eye. This is the rarest of our British species, but sometimes occurs in numbers on the coasts of Devon and Cornwall; it ranges from Scandinavia to the Mediterranean and thence southwards to the Congo. It is not known to grow to more than half the length attained by the Thick-lipped Mullet.

THE THIN-LIPPED GREY MULLET (*Mugil capito*)
(Pl. XXXIV, Fig. *b*) has the upper lip still thinner;
the maxillaries are exposed when the mouth is
closed, although not to the same extent as in the
Thick-lipped Mullet, but the rami of the lower jaw
are separated by an elliptical area; the pectoral fin
is shorter than in the other species, measuring less
than three-fourths of the length of the head, whilst
above its axil appears a scale which is more or less
elongate and is free except at the base. This is
the most widely distributed of our Grey Mullets,
extending from Scandinavia to the Cape of Good
Hope; it attains a length of 2 feet.

The Grey Mullets swim in shoals near the coasts,
and especially in the summer-time frequent estuaries,
going in and out with the tide, whilst at times they
may ascend far beyond tidal limits. Thus Yarrell
mentions that in the summer of 1834 Grey Mullets
migrated up the Arun in Sussex to 10 miles above
Arundel, whilst Thompson records that they have
been known to ascend the Lagan at Belfast into
the canal, where they have been shut in by the
gates and have been seen leaping in the fresh water.

They are very lively and often sport at the
surface; their food consists mainly of minute shellfish and of decomposed organic matter which they
extract from mud; their gills are protected from the
latter by a sieve-like apparatus formed by the
numerous gill-rakers. They are bold and cunning
and often escape the fishermen either by leaping
over or forcing a way under the nets; they are
sometimes captured by anglers with an artificial fly
or with a bait, but as they suck at the latter and
eject it immediately they have taken it into their

PLATE XXXIV

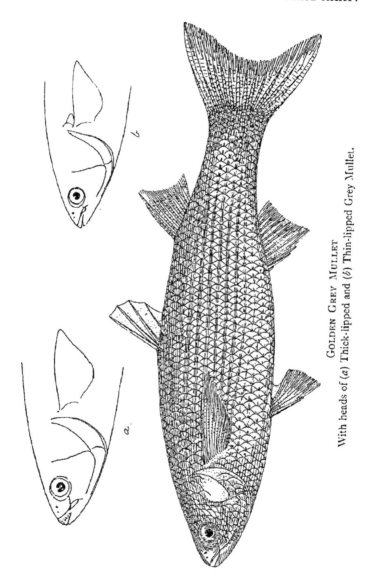

GOLDEN GREY MULLET
With heads of (a) Thick-lipped and (b) Thin-lipped Grey Mullet.

mouths, they are very difficult to hook; moreover, as they plunge and leap, rivalling the Trout in gameness, the hook is often torn from their tender mouths.

The Thick-lipped Grey Mullet spawns about May, usually in the sea; in July and August the fry, about three-fourths of an inch long, may be seen swimming in shoals at the surface; some of these were placed in the aquarium at Plymouth, and by the following August had attained a length of $2\frac{1}{4}$ to 3 inches.

All the Grey Mullets are held in high estimation as food, especially in continental countries; they will thrive and fatten in freshwater ponds, and are said to be improved by this treatment.

CHAPTER XII

THE BULL-HEAD AND THE STICKLEBACKS

THE Scorpænoid group differs from the Percoids in a single feature, namely, the prolongation of the second suborbital bone across the cheek to or towards the præopercle. Slight as this difference may seem, it is, nevertheless, very important, as with one exception this feature persists in all the numerous members of the suborder, including such diverse types as the Rock-perches (*Scorpænidæ*), Gurnards (*Triglidæ*), Flying Gurnards (*Dactylopteridæ*), Lump-suckers (*Cyclopteridæ*), and Bull-heads (*Cottidæ*); the single exception is the aberrant *Comephorus* from the depths of Lake Baikal, which has a very feebly ossified skeleton.

Plate XXXV

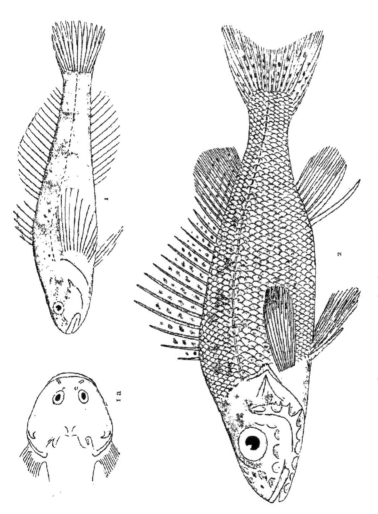

1. BULLHEAD; 1a. head seen from above; 2. RUFFE

Nearly all the Scorpænoids are marine, and only a small proportion of the Bull-heads occur in fresh water. The genus *Cottus* includes a considerable number of species from northern seas and rivers; of two European freshwater species one is found in our islands.

THE BULL-HEAD or MILLER'S THUMB (*Cottus gobio*) is a scaleless fish, with the head broad, depressed, and rounded in front, and the body subcylindrical, tapering backwards. The mouth is terminal, rather wide, with thickish lips, and with small pointed teeth forming bands in the jaws and across the head of the vomer. The eyes are rather small and are situated in the anterior half of the head and nearly on its upper surface. All the bones of the head are hidden beneath the skin, from which there emerges on each side a rather strong, curved, backwardly directed spine, arising from the præopercle; the gill-openings are restricted by the union of the gill-membranes with the isthmus. The spinous dorsal fin is short and low, composed of from six to eight simple, slender, flexible rays, which are unjointed, and are therefore regarded as spines in spite of the fact that they are not strong and pungent; it is followed immediately by the longer and higher soft dorsal fin, formed of fifteen to eighteen articulated rays which are either simple or else bifurcated at the tip; the anal fin is similar, but a little shorter, of eleven to fourteen rays; the caudal has the posterior edge truncate or slightly convex. The pectoral fins are large, with a broad oblique base, and the pelvic fins, of a small spine and three or four simple articulated rays, are inserted

18

below them.   The coloration of the upper surface is brownish or yellowish olive, spotted or marbled with dark brown or black, some of the spots usually uniting to form transverse bars ; the lower parts are yellowish white and the dorsal, caudal, and pectoral fins are barred with series of dark spots.

The Bull-head is found all over Europe except the Iberian Peninsula and Greece ; it extends into Russian Turkestan, but is replaced by an allied species in Siberia.   In our islands it occurs in England and Wales, but seems to be absent from Scotland and Ireland.   The example figured (Pl. XXXV, Fig. 1) is shown of the natural size ; it was sent to me by the late Dr. Bowdler Sharpe, who caught it in the brook at the bottom of his garden at Selborne.

This little fish is usually not more than 3 or 4 inches long, and 6 inches is the maximum length attained.   It is common enough, preferring clear brooks or shallow parts of lakes where the bottom is sandy or gravelly.   It lives a solitary life, and lurks on the bottom or in concealment under stones, waiting for the appearance of worms, insects, shrimps, small fry, etc., whereon it feeds ; although sluggish as a rule, and unable to swim for any distance at a time, it dashes out with great rapidity at its prey, and when disturbed shows a fine turn of speed in gaining the nearest available shelter.   It is a gluttonous fish, and if the opportunity occurs of seizing a prey of its own size, such as a Gudgeon or a Minnow, it will not hesitate to do so and to devour the victim at leisure.

The Bull-heads spawn in March or April, when they pair and select a shallow place in running water; here some sort of nest is prepared, usually a hole scooped out beneath a stone; the eggs are comparatively large and few, less than a thousand; they adhere together and are usually attached to the under surface of the stone which roofs the nest. The male guards the nest for about a month, keeping off intruders and protecting the eggs and young fry; the latter gather into a little shoal outside the nest and then separate and swim away, living thenceforth a solitary life until after two years they are mature and seek a mate

The Bull-head is able to inflict wounds with its spines, and from its habit of lying still in shallow water, either in concealment or in places where its colour renders it inconspicuous, it usually escapes the attentions of birds or fish of prey. Moreover, the width of the head, to which must be added the extended spines, makes it a difficult fish to swallow, and in 1880 a Grebe which had attempted this feat was found dead in the Isis.

The Bull-head is very remarkable for the rapidity with which it changes colour, the general hue of the body becoming pale or dark, the bars and spots appearing or disappearing, not only according to the environment, but under the influence of greed, fear, or anger.

It is very tenacious of life and lives for some time out of the water; although it is said to be well flavoured as food, and a good bait for Pike, it does not appear to be appreciated for either purpose in this country; nor is it worth the attention of the angler, although, as Walton says, " he never refuses to bite, nor indeed to be caught with the worst of anglers."

THE STICKLEBACKS (*Gastrosteidæ*) are the smallest of the British freshwater fishes, but they are by no means the least interesting. They are well worthy of attention on account of their pugnacity and their remarkable breeding habits, but still more for their great variability, which has been the cause of very divergent opinions as to the number of species which should be recognized.

The Sticklebacks form a well-marked group of fishes, the Gastrosteoids, which resemble the Scorpænoids in that the second suborbital bone is produced across the check to the præopercle, but differ from them precisely as the Grey Mullets differ from the Perches, in the loss of the attachment of the pelvic bones to the clavicles. The development on the chest of a pair of bony plates (*ectocoracoids*), which in the adult fish are completely united with the shoulder girdle, is highly characteristic.

The Stickleback family (*Gastrosteidæ*) comprises five genera, each with one, or at most a few, species. Two of these are North American, and a third, the Fifteen-spined Stickleback (*Spinachia spinachia*), although British, is exclusively marine. Our species are the Three-spined Stickleback (*Gastrosteus aculeatus*) and the Ten-spined Stickleback (*Pygosteus pungitius*).

THE THREE-SPINED STICKLEBACK (*Gastrosteus aculeatus*) is also known as Prickleback or Tittlebat, and has a host of other local names, such as Jacksharp, Pricky, Stickling, etc.; it is a little fish, never attaining a greater length than 4 inches, which is found on the coasts and in the rivers of the arctic and temperate regions of the Northern

Hemisphere, being recorded from as far north as Greenland, Alaska, and Kamchatka, and ranging southwards to Japan, California, New Jersey, and Spain.

The Three-spined Stickleback has the body compressed and fusiform, with the head conical and the caudal peduncle slender; the mouth is small, terminal, and oblique, and the jaws are furnished with narrow bands of sharp teeth; the gill-openings are restricted to the sides. The dorsal fin is composed of three, rarely two or four, spines, the two first of which are free, and of nine to thirteen soft rays; the anal is similar to the soft dorsal, and has one spine and seven to eleven soft rays; each pelvic fin consists of a spine similar to those on the back, with a single soft ray in its axil, and is so constructed that when everted it can be fixed by means of an inner basal knob which catches against the pelvic bone; the pectoral fins have a vertical base, in front of which there is on each side a glistening naked area bounded below by the bony plates (*ectocoracoids*) mentioned above as peculiar to this group; behind this pair of chest plates the united pelvic bones appear, when seen from below, as a lanceolate, triangular, heart-shaped or V-shaped plate with the point directed backwards.

There are no true scales, but on the sides of the body a series of vertically expanded bony scutes is variously developed, in some examples extending from the head to the caudal fin, in others reduced to three or four in the region of the pectoral fin. The sixth plate (of the complete series) is usually joined below to an ascending lateral process of the pelvis, with which the fifth and seventh may also be in contact ; when reduction takes place these three plates (fifth to seventh) are the most persistent, whilst the first to disappear are those which precede the small keeled plates on the posterior part of the tail.

Some of the older writers recognized three distinct species according to the development of the plates,

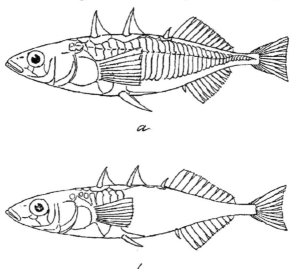

FIG. 22.—Three-spined Sticklebacks   a. Rough-tailed (*trachurus*) form.   *b* Smooth-tailed (*gymnurus*) form.

namely, the Rough-tailed Stickleback (*Gastrosteus trachurus*) with a complete series of plates from the head to the caudal fin, the Half-armed Stickleback (*G. semiarmatus*) with a naked interspace between the large anterior plates and the small keeled plates of the caudal peduncle, and the Smooth-tailed Stickleback (*G. gymnurus*) with the body naked except for a few anterior plates.

However, intermediate forms may occur, as for example specimens exhibiting the *semiarmatus* type on one side of the body and the *trachurus* type on the other, and the study of a large series of specimens shows that there is a continuous gradation from the *gymnurus* to the *trachurus* armature.

In the British Isles it may be generally stated that in inland localities all the Sticklebacks are of the *gymnurus* type; this applies to a number of examples from various parts of Ireland sent to me by Mr. Robert Patterson, to a series of specimens from Newark for which I am indebted to Mr. R. Littler, and to some from Shrewsbury, given to me by Mr. H. E. Forrest, as well as to many others from various parts of Britain. Sticklebacks from estuaries, or places not far from the sea, exhibit a remarkable inconstancy in the development of the plates, all gradations from the *gymnurus* to the *trachurus* type being often found in the same shoal. In the sea, at any rate on the northern coasts of our islands, the *trachurus* type prevails.

In the northern part of its range the Three-spined Stickleback is typically a marine species with strong dermal ossification, the series of bony plates complete and the individual plates deep, the caudal keel strong, the ectocoracoids long and the naked areas above them and in front of the pectoral fins consequently large, the pelvis long and lanceolate, without an anterior notch, and the fin-spines either long or strong. I have compared examples of this type from the Shetlands with others from Alaska and Puget Sound, and I am unable to detect any difference. In these high latitudes a reduction of the dermal ossification is found in freshwater

Sticklebacks; thus typical examples of the *gymnurus* form occur in the rivers of Greenland.

Towards the south the Three-spined Stickleback is more generally a freshwater fish, and there is reason to doubt whether it ever goes out to sea at all in the Mediterranean. In our islands, as already mentioned, a great variability is seen in specimens from localities near the sea; but when the lateral series of plates is continuous (*trachurus* form) they are not so deep and are often less numerous than in northern examples, the caudal keel is weaker, the naked area in front of the pectoral is smaller, the pelvis is shorter and is notched in front, the fin-spines are shorter or weaker, and the number of fin-rays is usually less. Examples from Wandsworth (*gymnurus, semiarmatus,* and *trachurus*) differ in no respect from others from the coast streams of California.

A nominal species, first described from Italy, is the so-called Short-spined Stickleback (*Gastrosteus brachycentrus*); this is only a *gymnurus* form with very short fin-spines, and may often be met with in inland localities in Ireland. I have compared specimens of the *gymnurus* form from various parts of the British Isles with others from Spain and Italy, and from such far-away places as California and Lake Biwa in Japan, without detecting any differences.

It is only when we get to the extreme southern part of the range of the Three-spined Stickleback, where it does not enter the sea, that we get freshwater colonies exhibiting peculiarities which may be regarded as specific. Thus I have described a form from Rome (*G. hologymnus*) in which the bony plates are entirely wanting and the snout is longer than in our Stickleback, whilst the Algerian Stickleback (*G. algeriensis*) may be shortly characterized as a *gymnurus* form with not more than four small plates on each side, the spines very small, the pelvis very short and broad, and the vertebræ reduced in number, only twenty-nine instead of thirty-one to thirty-three

Four-spined examples are not very uncommon, and have been wrongly described as distinct species ; a series from Newark sent to me by Mr. Littler includes not only a four-spined specimen but also a two-spined one, the latter being a great rarity.

As the Three-spined Stickleback is found both in fresh and salt water, some curious experiments have been made to ascertain the result of sudden transference from one medium to the other. These seem to show that, as might be expected, when Sticklebacks accustomed to live in fresh water are suddenly placed in salt water they do not long survive the change, on the other hand, estuarine examples of any of the varieties can be readily transferred suddenly from salt to fresh water and *vice versa*, without seeming to notice the difference.

The Three-spined Stickleback lives in shoals and is especially abundant in small streams, ditches, and ponds. The shoals are sometimes very large, and as an example of the extraordinary numbers attained we may quote Pennant, who in 1776 wrote : " Once in seven or eight years amazing shoals appear in the Welland and come up the river in the form of a vast column The quantity is so great that they are used to manure the land, and trials have been made to get oil from them. A notion may be had of this vast shoal by saying that a man employed by the farmer to take them has got for a considerable time four shillings a day by selling them at a halfpenny a bushel."

A similar phenomenon was observed in April, 1909, at Whittlesey, for Mr. J. E. Rowell wrote to the *Fishing Gazette* that vast hordes of Three-spined Sticklebacks had lately shown themselves in the rivers and canals in the neighbourhood, although previous to this visitation they were scarcely known in the district. Probably these enormous migrating shoals come in from the sea.

According to Smitt, in the late autumn the Three-spined Sticklebacks roam about in large companies and yield a rich harvest to the Baltic fishermen, who catch them by means of seine-nets, or in the evening attract them with torchlights and use hand-nets for their capture ; they are boiled down into oil and the sediment is used as manure. The author just quoted also states that the fish is hardly used at all as human food, but that in England, together with Herring fry, it often tempts consumers under the name of Whitebait.[1] In North America and Kamchatka Sticklebacks are used as manure and as food for cattle.

Although so insignificant in size, the Three-spined Stickleback is bold and greedy, often fiercely attacking larger fishes ; it feeds on shrimps, insects, worms, etc., and is especially destructive to the spawn and young fry of other fish.

---

[1] An analysis of Whitebait shows that there is usually a considerable proportion of the fry of Herrings and Sprats ; the other constituents vary greatly according to locality and season, but Smelts, Sand-smelts, Sand-eels, and Sticklebacks are often present.

Day gives an account, which has often been quoted, of how some Roach were introduced into an aquarium containing some Three-spined Sticklebacks; the prior inhabitants were dissatisfied and forthwith attacked the new-comers, continuing until all the Roach had been killed, when they were eaten by their conquerors. Baker observed that in five hours a Stickleback devoured seventy-four young Dace, each about a quarter of an inch long, and that two days afterwards it swallowed sixty-two. The pertinacity with which they hold on to anything they may have seized often causes their downfall, children angling successfully with a stick and a piece of thread to which a worm is tied.

The Three-spined Stickleback breeds in the spring or summer, the time of spawning varying considerably in different years and according to the locality. When the breeding season commences the coloration begins to change, the dark greenish colour of the back extending on to the sides in the form of vertical bars, whilst the lower parts change from a silvery white to a pale yellow in the female and a brilliant red in the male. The males, who seem to be much fewer than the females, select suitable places for constructing their nests, such as quiet shallows, or rock pools which are only reached by the sea at high tides; each chooses a spot at some little distance from his neighbours, and commences the building of a nest, first collecting together pieces of the roots and stalks of aquatic plants and arranging these materials on the bottom, and then building them up and cementing them together by means of threads, which are of the nature of a mucous secretion of the kidneys. The nest is made carefully, and is attached to the bottom by means of sand and pebbles; its completion takes several days, and when finished it is usually a barrel-shaped structure, often an inch or more in diameter and with an opening at one end; sometimes it is dome-shaped, with the aperture at the top.

So far the fish has been intent upon his work, only trying to provide against any interruption to its progress and watching with suspicion every creature that approaches; now he devotes his energies to attracting a mate, whom he guides to the nest, swimming round her with evident pleasure and driving her in by thrusting at her with his snout, or, if she be unwilling, even using his spines. When the female has entered she lays a few eggs and then bores a hole through the wall of the nest on the side opposite to the entrance, and departs; immediately after the eggs are laid the male enters the nest and fertilizes them. On the following day he goes in search of another spouse, and this is repeated until the eggs are sufficiently numerous.

At this time of courtship and matrimony the males are violently jealous and often indulge in fierce duels, the two rivals rushing at each other, dealing violent strokes with their pelvic spines, and then hastily returning to the neighbourhood of their nests, after a few rounds one gives in and then the victor indulges in a splendid display of colours, whilst those of the vanquished become dull; these fights are often mortal, one combatant frequently ripping his opponent open with his sharp spines.

When the spawning is over the male guards and protects the eggs, furiously attacking any other Stickleback that may approach; with his mouth he repairs every damage to the nest and often renews the water by placing himself in front of the opening and vibrating his pectoral fins. When the eggs hatch out he destroys the nests, leaving only the foundations as a sort of cradle, which he continues to guard; as the young gain in size and strength they try to leave the nest, but are always intercepted by the father, who takes them in his mouth and returns them to their quarters. At last his care begins to slacken, and finally they are left to shift for themselves.

THE TEN-SPINED STICKLEBACK or TINKER (*Pygosteus pungitius*) differs from the preceding

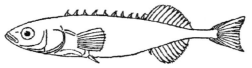

FIG. 23.—Ten-spined Stickleback.

species especially in the more numerous dorsal fin-spines, which vary in number from seven to twelve, and are alternately divergent to the right or left, and in the wider gill-openings, which are confluent below. It is a smaller fish than the Three-spined Stickleback, attaining a maximum length of 3 inches. It has nearly the same geographical distribution as the Three-spined Stickleback, but does not extend so far south, in Europe not crossing the Alps; in our islands there are no reliable records from Scotland north of Loch Lomond and the Forth.

Except in the far north the Ten-spined Stickleback is principally a freshwater fish, but is found in the Baltic; until a study has been made of a much

larger series of specimens than I have seen it will be
impossible to say how many races or subspecies are
definable or whether any of them deserve to rank
as distinct species. The number of fin-rays varies
considerably in different localities, and so does the
development of the lateral plates ; one of the most
remarkable forms is that which inhabits the tributaries
of the Black and Caspian Seas (*P. platygaster*), which
resembles the Three-spined Stickleback in that the
fourth to the seventh plates of the complete series,
near the ascending process of the pelvis, are the
most developed, whereas Ten-spined Sticklebacks
from other parts of the world have the anterior
plates not much deeper than the others when the
series is complete, whilst the posterior plates are the
most persistent.

In Western Europe the Ten-spined Sticklebacks
are either entirely naked or have a few small plates
bearing a keel on each side of the caudal peduncle ;
all the British and Irish specimens I have seen agree
with those of France and Belgium in having the
spines much smaller than in those from Scandinavia ;
the correct name for this short-spined form, if it be
recognized as a distinct race, is *Pygosteus lævis*.

In examples from various parts of the British
Isles I count eight to eleven spines and nine to
twelve soft rays in the dorsal fin, and one spine and
eight to eleven soft rays in the anal. Specimens
entirely lacking the pelvis and pelvic fins have been
taken at Tipperary and Edgeworthstown in Ireland ;
of thirteen captured by Day at the latter place eleven
showed this peculiarity.

The coloration of the Ten-spined Stickleback
varies considerably, but is usually a greenish olive,

powdered with little black dots which may be arranged to form irregular bars; in the breeding season the males are dark brownish.

In habits this species is rather similar to the preceding one; it spawns in small streams, where the shoals crowd in close to the banks, and the nests are not built on the bottom, but are attached to weeds or grasses.

CHAPTER XIII

THE FLOUNDER

Peculiarities of Flat-fishes—Percoid characters of *Psettodes*—distribution of Pleuronectids—the Flounder described—coloration—differences from the Plaice—reversed examples—ambicolorate specimens—albinos—distribution—habits—breeding—growth and transformation—value as food

WE have already mentioned that the Perches represent the highest stage in the evolution of the normal fish, but that from them various groups have evolved in adaptation to special modes of life. Such a group is that of the Flat-fishes (*Pleuronectidæ*), which form a well-marked order which has received the name *Heterosomata*.

These fishes have a strongly compressed body and differ from all other fishes in that they lie on one side at the bottom of the sea; consequently only the exposed side is coloured in such a way as to render the fish inconspicuous, whilst the hidden under-side is white. Still more remarkable is the shifting of the eye of the under-side from its original position, where it is of no use, round into the light, so that both eyes appear to be on the same side of the head; thus, in Flat-fishes we can distinguish between an eyed or coloured side and a blind or white side.

In most Pleuronectids the fins are composed of articulated rays only; the dorsal and anal are long and fringe the dorsal and ventral edges of the flat body; the caudal has the normal homocercal structure and the pelvics, formed of six or fewer rays, are placed in advance of the pectorals. In the adult fish the air-bladder is absent; as in the Percoid fishes the præmaxillaries form the entire upper border of the mouth, and another Perch-like feature is the firm and direct attachment of the pelvic bones to the clavicles.

The most generalized of all the Flat-fishes is the one named *Psettodes erumei*, found on the coasts of Africa, India, and China. In this species sinistral and dextral individuals are equally numerous, and it differs from all other Pleuronectids in that the eye of the blind side comes on to the top of the head only, whilst the dorsal fin does not extend forward on to the head, and anteriorly is formed of slender spines; moreover, each pelvic fin is formed of a spine and five soft rays, as in a Perch. The wide mouth and strong sharp teeth give the clue to the retention of these Percoid features; this is evidently an active and predaceous fish, often swimming by strokes of the tail. It is in the Flat-fishes which usually progress quietly by undulating movements of their fringing fins that the spines have become reconverted into articulated rays and the dorsal fin has grown forwards on to the head, and the more they have become accustomed to living and feeding on the bottom the more has their asymmetry affected not only the eyes and skull, but the mouth and dentition, the paired fins, and the position of the vent.

19

The Flat-fishes inhabit the seas of all parts of the world, except the polar regions; some are found in deep water, but the majority prefer sandy shores; quite a number ascend rivers for some distance, but in our islands only one, the Flounder, constantly occurs in fresh water above tidal limits.

THE FLOUNDER or FLUKE (*Pleuronectes flesus*) has the body ovate in form, with the greatest depth nearly one-half the length. The eyes are usually placed on the right side and are separated by a bony keel, which is continued backwards as a smooth or tuberculated postorbital ridge to the origin of the lateral line. The mouth is rather small, terminal, oblique, with the lower jaw a little projecting; the teeth are pointed in the young, but in the adult they are chisel-shaped and arranged in a single series; on the blind side the jaw-bones are stronger and the teeth more numerous; for example, in a specimen about a foot long I count on the blind side twenty-two teeth in the upper jaw and the same number in the lower, but on the eyed side only ten upper and thirteen lower teeth.

The dorsal fin commences above the upper eye and is composed of from fifty-three to sixty-three simple rays, the middle ones being the longest; the anal is similar, of thirty-seven to forty-five rays, commencing not far behind the pelvics, and preceded by a forwardly directed spine; the caudal fin is well developed, truncate, borne on a peduncle which is nearly as long as deep; the pectoral and the six-rayed pelvic fins are equally developed on both sides. The body is covered with small, smooth scales, whilst modified scales in the

PLATE XXXVI

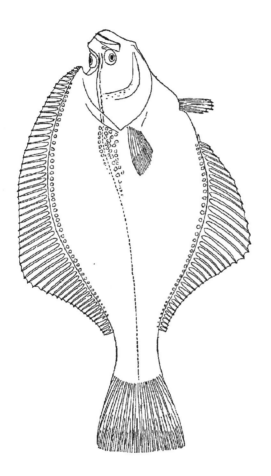

FLOUNDER (ambicolorate specimen)

form of spinous tubercles are also present, usually forming a group near the origin of the lateral line and a series along the bases of the dorsal and anal fins, corresponding to the interspaces between the rays ; the lateral line is nearly straight, with a slight curve above the pectoral fin.

Like most Flat-fishes the Flounder is able to change its colour according to the nature of the ground, and the resemblance to the sand, gravel, or mud on which it lies is so perfect that it is almost impossible to see the fish unless it moves. In this species it has been definitely established that the contractions and expansions of the pigment cells are under the control of the nervous system, and that the perception by the eye of the colour of the bottom is necessary for the assumption of a similar appearance by the fish.

The colour is often greyish olive, occasionally marbled with brownish ; but it varies from nearly yellow to almost black ; orange spots like those of the Plaice are sometimes present. The latter fish is the nearest ally of the Flounder, from which it is especially distinguished by the more numerous dorsal and anal rays, the shorter caudal peduncle, the absence of spinous tubercles on the body, and the presence of bony knobs on the postorbital ridge.

Reversed examples of the Flounder, *i e.* specimens with the eyes and colour on the left side, are fairly common, and in some localities are at times as numerous as the others.

Ambicolorate Flounders are by no means rare, and it is interesting to note that many of these are not only symmetrical in having both sides coloured alike, but are less asymmetrical than usual in that the upper eye has come only on to the upper edge of the head, so that the anterior extension of the dorsal fin has been unable to join the head above the eye, as is shown in the example figured (Pl. XXXVI). Albinos, quite white and with pink eyes, have sometimes been captured, and also partial albinos, rosy, with a few dark spots.

The Flounder attains a length of about 18 inches ; it is common all round Europe. During the summer it lives in shallow water near the shores, especially where the bottom is sandy ; it also frequents estuaries, and in our islands ascends all suitable rivers, usually as far as the first falls which bar its progress. Thus it formerly ascended the Severn to Shrewsbury, but is now unable to get beyond the weirs which have been constructed near Gloucester. It is a sluggish fish, usually lying partly buried in the sand, sometimes with only the eyes exposed ; when it swims it progresses chiefly by means of undulating movements of the dorsal and anal fins ; but it will often take a jump from one place to another, if the expression is permissible when the fish does not leave the water, lifting itself from the bottom by a stroke of the tail. It feeds on worms, little fish, etc., but especially on crustaceans and molluscs, the pharyngeal teeth being strong and obtuse, well adapted for crushing shells.

The spawning season is usually in March and April, but varies from January to June ; at this season the Flounders assemble in the sea in moderately deep water, usually between 5 and 30 fathoms ; the eggs are very numerous ; they are buoyant and float at the surface, and are hatched in about a week ; the larvæ are symmetrical and moderately elongate in form, and when they are about a week old, and one-sixth of an inch long, the yolk is all absorbed, the mouth has formed, and they commence feeding on minute organisms; after a month or more, when they are less than half an inch long, they are essentially similar in organization to the adult fish, except that they are not so deep in the body, and are still symmetrical; now they sink to the bottom and lie on one side, and the eye which is underneath moves to the edge of the head and then round on to the upper side; at the same time the dorsal fin is prolonged forward, and as soon as the eye has come round grows along the edge of the head above it. In some species of Flat-fishes this forward growth of the dorsal fin precedes the migration of the eye, which then has to push its way between the base of the fin and the edge of the head, and so actually seems to travel through from one side of the head to the other. During their transformation the little Flounders approach the shore and enter the bays and estuaries; about May specimens half an inch long may often be found in tidal pools. The rate of growth varies enormously, but they become sexually mature when two or three years old and not less than 6 inches long.

Although very abundant, the Flounder is generally not held in much estimation for the table, and is considered much inferior to its relative, the Plaice. It may be captured by the angler, who thinks it worth the trouble, by using a small worm as a bait.

CHAPTER XIV

BRITISH FRESHWATER FISHES—THEIR ORIGIN AND GEOGRAPHICAL DISTRIBUTION

Origin of freshwater fishes—some still spend part of their life in the sea. Catadromous fishes. Anadromous fishes. Freshwater colonies of marine species—Char and Whitefish—true freshwater fishes, belonging to freshwater families. Zoogeographical divisions. Distribution of British species—poverty of Irish fish-fauna—decrease in number of species northwards and westwards in Britain—our species found on the Continent. Recent geological history of the British Isles. Establishment of colonies of Char and Whitefish. Union of Britain with the Continent—our true freshwater fishes come from the Rhine and the Seine—means of dispersal in the British Isles.

NO doubt the first fishes who left their original home in the sea and entered fresh water did so in pursuit of food, or to escape large and predaceous enemies, or to seek quiet places where they could breed unmolested. Some of these visitors became permanent residents, and so formed freshwater races and, in time, species distinct from their marine allies ; in the course of ages these spread and became further diversified, and so it happens that at the present day there are families, and even orders, composed entirely of freshwater fishes. There is no special character common to freshwater fishes ; nearly every order has some freshwater representatives, and they have been moulded by their manner of life into the most diverse forms.

Our freshwater fishes may be primarily divided into (*a*) those which spend a part of their life in the sea and (*b*) permanent residents in fresh water. In the former class are included marine fishes such as the Grey Mullets and the Bass, which frequent estuaries and are occasionally found higher up the rivers, and the Flounder, which migrates up stream for considerable distances. The Three-spined Stickleback, which occurs on all our coasts and in nearly all our rivers, is equally at home in the sea and in fresh water, and breeds in either; in the Arctic regions it is essentially a marine fish, but towards the southern limits of its area, in Spain and Italy, it rarely enters the sea. Catadromous fishes, which descend to the sea to breed, are represented in our islands by the Eel, which returns to its ancestral home in the depths of the ocean for the purposes of reproduction; the abundance of Eels in our waters is due to our proximity to places of the right depth and temperature for their breeding, in the Atlantic to the west of Ireland. Anadromous fishes are those which ascend from the sea to spawn in fresh water; under suitable conditions these may lose their migratory habit and may found colonies of permanent freshwater residents. The Salmon, the Shads, and the Sea Lamprey, when adult, feed and grow in the sea, and enter our rivers only in order to spawn. But it is interesting to note that Lake Wenern in Sweden, and some of the larger lakes and rivers of Quebec, New Brunswick, and Maine are inhabited by Salmon which never go to the sea; it is exactly the same with the Sea Lampreys of some of the lakes of the State of New York ; in times past the abundance of food in the lakes has induced them to give up their journey to the sea, and if nowadays they could get there they would certainly not be able to get back again, and so they remain, freshwater colonies or races of anadromous species.

The Trout and the River Lamprey, which are respectively somewhat smaller species than the Salmon and the Sea Lamprey, differ from the latter also in that they feed and grow in fresh water as well as in the sea, and in all suitable rivers and lakes which they can reach a proportion of them will become permanent freshwater residents.

From a consideration of those fishes which spend part of their life in the sea we have thus gradually arrived at those which pass the whole of their lives in fresh water. Fluviatile colonies of the Three-spined Stickleback and the Trout may often be reinforced from the sea, and in our islands only show peculiarities immediately associated with their changed environment, but the Char and Whitefish, long isolated in their lakes, are very different from their marine allies.

The range in the sea of a migratory species such as the Salmon seems to be mainly a matter of temperature; there is a pretty definite southern limit, and we may take it that if the climate of Europe were colder this limit would be farther south. Thus we may explain the presence of a Three-spined Stickleback and a Trout in the rivers of Algeria, farther south than the present range of these fish in the sea, on the hypothesis that they reached those rivers from the sea when the climate was colder, and we may suppose that their isolation from the parent stock has led to the assumption of peculiarities which enable them to be recognized as distinct.

The only species or races of freshwater fishes peculiar to the British Isles have had a similar history; they are the Char (*Salvelinus*) and Whitefish (*Coregonus*), fishes of the Salmon family. In our islands, as in Scandinavia and in the Alps, Char are now isolated in various lakes, and never go to the sea, but these lacustrine forms of Char are closely related to anadromous Arctic species; our Whitefish also are allied to northern migratory species. There can be little doubt that the Char and Whitefish reached the lakes which they now inhabit from the sea when the climate was colder, as we know it must have been in comparatively recent times, and that during their long isolation they have evolved in various directions, and at varying rates according to circumstances, so that several distinct forms may now be recognized.

The following list includes the

FRESHWATER FISHES PECULIAR TO THE BRITISH ISLES

Species	*Distribution*
1. Windermere Char (*Salvelinus willughbii*, Gthr.).	Windermere, Coniston, Wast Water, Ennerdale, Crummock Water, etc., also represented by very similar forms in many Scottish lakes.
2. Torgoch or Welsh Char (*S. perisii*, Gthr.).	Lakes of Llanberis and some other lakes in Carnarvonshire and Merionethshire.
3. Lonsdale's Char (*S. lonsdalii*, Regan)	Haweswater, Westmorland.
4 Struan (*S. struanensis*, Maitland).	Loch Rannoch, Perthshire.

Species	*Distribution*
5 Haddy or Killin Char (*S. killinensis*, Gthr.).	Loch Killin, Inverness-shire, represented by a very similar form in Loch Roy, Inverness-shire.
6. Malloch's Char (*S. mallochi*, Regan).	Loch Scourie, Sutherlandshire.
7 Large - mouthed Char (*S. maxillaris*, Regan).	Loch under Ben Hope, Sutherlandshire.
8. Orkney Char (*S. inframundus*, Regan).	Formerly in Hellyal Lake, Hoy Islands, Orkneys, now probably extinct.
9. Shetland Char (*S. gracillimus*, Regan)	Loch of Girlsta, near Lerwick, Shetlands.
10. Cole's Char (*S. colii*, Gthr.)	Loughs Eask and Derg in Donegal, Conn, Mask, Corrib, Inagh, etc., in Galway, Gortyglass in Clare, and Currane in Kerry; the Irish representative of the Windermere Char.
11. Trevelyan's Char (*S. trevelyani*, Regan)	Lough Finn in Donegal.
12. Gray's Char (*S. grayi*, Gthr).	Lough Melvin in Fermanagh.
13. Scharff's Char (*S. scharffi*, Regan).	Lough Owel in Westmeath.
14. Coomasahan Char (*S. fimbriatus*, Regan).	Lough Coomasaharn in Kerry
15. Blunt-snouted Irish Char (*S. obtusus*, Regan).	Loughs Luggala and Dan in Wicklow, Acoose and Killarney in Kerry.
16. Lochmaben Vendace (*Coregonus Vandesius*, Richards.).	Castle and Mill Lochs, Lochmaben, Dumfriesshire.
17. Cumberland Vendace (*C. gracilior*, Regan).	Derwentwater and Bassenthwaite.
18. Lough Neagh Pollan (*C. pollan*, Thomps)	Lough Neagh.
19. Lough Erne Pollan (*C altior*, Regan).	Lough Erne.

Species.	Distribution
20. Shannon Pollan (*C. elegans*, Thomps.)	Lakes of the Shannon System.
21. Powan (*C. clupeoides*, Lacep).	Lochs Lomond and Eck.
22. Schelly (*C. stigmaticus*, Regan).	Haweswater, Ullswater, and the Red Tarn.
23. Gwyniad (*C. pennantii*, Cuv. and Val.).	Bala Lake, Merionethshire.

The Char are essentially fishes of mountain lakes, which are usually deep and cold ; in our islands they are found in Scotland, Ireland, the Lake District of England, and North Wales, in fact, in all parts where there are suitable lakes

The Whitefish belong to three types : the Vendaces, inhabiting lakes connected with the Solway, are two closely related forms which are quite distinct from their continental allies in the countries round the Baltic, species of the Vendace type also ascend Siberian rivers from the Arctic seas. The three forms of Pollan in the Irish lakes are also very similar to each other, but are not represented in Europe, although Pollan-like fishes inhabit the Arctic Ocean and run up the rivers of Siberia The Powan, Schelly, and Gwyniad are only local forms of one species, which is closely related to *Coregonus wartmanni* of the Alps, to forms inhabiting Scandinavia, and to anadromous Arctic species.

So far we have dealt only with fishes which spend a part of their life in the sea, or which have been quite recently derived from anadromous species, the latter category including all the forms peculiar to the British Isles. We have now to deal with fishes which pass the whole of their lives in fresh water, which are identical with continental species, and which came to us at the time of our last union with the Continent. For the most part these belong to families which consist wholly of freshwater fishes,[1] which have been long established and have evolved their genera and species in fresh water; such families are of the greatest importance in the solution of the problems of geographical distribution, especially in pointing out the existence of former land connexions or of ancient lines of severance.

We may here indicate that the freshwater fishes call for a primary separation of the world into two main zoogeographical divisions, on the one hand America, Africa, Europe, and Asia, including the Philippines, Java, Sumatra, Borneo, and Bali, on the other New Zealand and Australia, with New Guinea, Celebes, and Lombok. Wallace's Line, drawn between Borneo and Celebes and Bali and Lombok, probably represents the line of the final separation of the Asiatic and Australian land-masses at or before the beginning of the Tertiary epoch. The true freshwater fishes, which have evolved in the rest of the world, have never found their way into the Australian region, wherein, except for two persistent archaic types, the freshwater fishes are closely allied to the marine ones; peculiar genera and species of Gobies, Sea-perches, Grey Mullets, Sand-smelts, etc., are plentiful in the rivers of Australia and New Guinea, but there are no peculiar families confined to fresh waters in this area. The contrast between Celebes, without a

[1] Many species, of course, can stand brackish water; Roach, Perch, etc., occur in parts of the Baltic and Caspian, where the salinity is very low. Species of Pike-perch (*Lucioperca*), fishes of the Perch family, even thrive in the Black Sea, where, however, the salinity is only half that of ordinary sea water.

single indigenous true freshwater fish, and the neighbouring island of Borneo, with its hundreds of species of Cat-fishes, Carps, Loaches, etc., is most striking.

Having thus eliminated the Australian region, the remainder of the world may be divided into five regions of unequal value, corresponding roughly to the continents of South America, North America, Europe (with Northern Asia), Africa, and Asia south of the central plateau. Of these, the Palæarctic

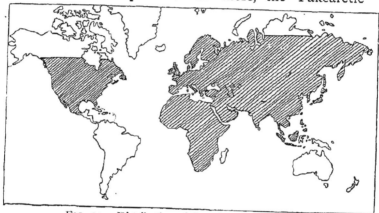

FIG. 24.—Distribution of Carp family (*Cyprinidæ*).

region, comprising Europe and Asia north of the Himalayas, exhibits considerable affinity with the Nearctic, or North American region, none at all with South America, very little with Africa, but more with tropical Asia; indeed, in Southern China the Palæarctic and Indian regions merge into each other. The Palæarctic region is characterized especially by the presence of Perches (*Percidæ*), Pikes (*Esocidæ*), Carps (*Cyprinidæ*), Loaches (*Cobitidæ*), etc., and by the absence of several families peculiar either to North America or to the tropical regions. Of the

families just mentioned the two first are essentially
northern, but the two last show their greatest variety
in the Indian region, which may be regarded as their
original home; all four of these are represented in
the British Isles.

The Grayling belongs to the Salmon family, but to
a genus (*Thymallus*) which is strictly fluviatile and
does not, like the other genera, include anadromous
species.   The Burbot is the only freshwater fish of

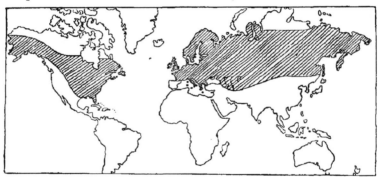

FIG. 25.—Distribution of Pike family (*Esocidæ*).

the Cod family, and is generically distinct from any
of the marine forms.   The Miller's Thumb and
Planer's Lamprey are freshwater species of genera
which are respectively marine and anadromous, and
our Ten-spined Stickleback is a freshwater race of a
species which is marine in the far north.   But the
evidence from their distribution and from their identity
with continental forms leads to the conclusion that,
so far as we are concerned, these are true freshwater
fishes which reached us from the Continent.   Thus
they may be added to the species of the four families
above mentioned in the following list of the

TRUE FRESHWATER FISHES OF THE BRITISH ISLES

Species	Distribution in British Isles	Distribution outside British Isles
Family PETROMYZONIDÆ— 1. Planer's Lamprey	Britain, except possibly the Northern Highlands of Scotland; Ireland.	Europe, Siberia, Japan.
Family SALMONIDÆ— 2. Grayling	England and East Wales (local); South Scotland (introduced), absent from Ireland	Northern and Central Europe, southward to Northern Italy.
Family CYPRINIDÆ— 3. Crucian Carp	England (local, perhaps introduced).	Europe, Russian Turkestan, Siberia, Mongolia.
4. Barbel	Thames, Trent, and some Yorkshire rivers.	France, Germany, Danube.
5. Gudgeon	England, except Lake District; Wales, except in the west; Ireland; absent from Scotland.	Europe, except Iberian Peninsula and Greece; Russian Turkestan, Siberia, Mongolia.
6. Tench	Britain, except Northern Highlands of Scotland; Ireland	Europe, Asia Minor, Western Siberia.
7. Roach	Britain, north to Loch Lomond and the Teith, absent from Cornwall and Ireland	Europe, north of the Alps and Pyrenees; Russian Turkestan, Siberia

20

TRUE FRESHWATER FISHES—*continued*

Species	Distribution in British Isles	Distribution outside British Isles
Family CYPRINIDÆ—*contd.* 8. Dace	England, Wales except in the west; absent from Scotland and Ireland.	Europe, north of the Alps and Pyrenees; Siberia
9. Chub	Britain, except West Wales, Devon, Cornwall, and Highlands of Scotland; absent from Ireland.	Europe, except Iberian Peninsula, Asia Minor, Persia.
10. Minnow	Britain, except Northern Highlands of Scotland; Ireland (local).	Europe, except Iberian Peninsula, Russian Turkestan, Siberia.
11. Rudd	England and Wales (rather local); Ireland; absent from Scotland.	Europe, except Iberian Peninsula; Asia Minor, Russian Turkestan, Western Siberia.
12. White Bream	Eastern rivers from Yorkshire to Suffolk.	Europe, north of the Alps and Pyrenees; Western Siberia.
13 Bream	Britain, except Highlands of Scotland and Dorset, Devon, and Cornwall; Ireland.	Europe, north of the Alps and Pyrenees; Turkestan, Western Siberia.
14. Bleak	England, except Lake District and south-west counties; Wales, except in the west; absent from Scotland and Ireland.	Europe, north of the Alps and Pyrenees.

Family COBITIDÆ— 15. Loach	Britain, except Northern Highlands of Scotland; Ireland.	Europe, except the Iberian Peninsula and Greece; Siberia, Mongolia, China, Japan.
16. Spined Loach	England (rare and local); probably absent from Wales, Scotland, and Ireland.	Europe, Russian Turkestan, Siberia, Mongolia, China, Japan.
Family ESOCIDÆ— 17. Pike	Britain (local in north of Scotland); Ireland.	Europe, except the Iberian Peninsula; Russian Turkestan, Siberia, Mongolia; North America, region of the Great Lakes and north-west to Alaska.
Family GADIDÆ— 18. Burbot	Eastern rivers from Durham to Suffolk.	Europe, except the Iberian Peninsula and Greece; Siberia; North America, region of the Great Lakes and northwards.
Family GASTROSTEIDÆ— 19. Ten-spined Stickleback (short-spined race)	Britain, north to Loch Lomond and the Forth area; Ireland	France, Belgium, Germany (?).
Family PERCIDÆ— 20. Perch	Britain, except Northern Highlands of Scotland; Ireland.	Europe, except the Iberian Peninsula, Russian Turkestan, Siberia
21. Ruffe	England, except Lake District and south-west counties; Wales, except in the west; absent from Scotland and Ireland.	Europe, north of the Alps and the Pyrenees; Russian Turkestan, Siberia.
Family COTTIDÆ— 22. River Bull-head.	England and Wales; absent from Scotland and Ireland.	Europe, except the Iberian Peninsula and Greece; Russian Turkestan.

From this it will be seen that all of the twenty-two species are inhabitants of the continent of Europe, and that whilst only one of them crosses the Pyrenees into Spain, and about half of them get over the Alps into Italy, quite a number extend far into Asia, and two even are common to Europe and North America. Moreover, it will appear that whilst all the species occur in Yorkshire, and nearly all in the Trent, the Ouse, and Norfolk, there are parts of these islands where the true freshwater fish-fauna is a very poor one. Ireland has only ten of the twenty-two species, namely, Planer's Lamprey, Gudgeon, Tench, Minnow, Rudd, Bream, Loach, Pike, Ten-spined Stickleback, and Perch, and one of these, the Minnow, is reputed to have been introduced about one hundred years ago. In Britain there is a considerable diminution in the number of species towards the north. Thus in Loch Lomond and the Teith there are present only Planer's Lamprey, Tench, Minnow, Roach, Loach, Pike, Ten-spined Stickleback, and Perch. Of these the Roach is absent from the Tay, Dee, and Deveron, whilst the Tench occurs in these systems only in ponds where it has been introduced. Finally, in the Northern Highlands, as in the outlying islands, there seem to be no indigenous true freshwater fishes, although the Pike has been introduced into some of the lochs and rivers, and the Minnow into the Spey.

A similar decrease in the number of species takes places from east to west; the Burbot, the Barbel, and the White Bream are found only in English rivers which flow into the North Sea. In Dorsetshire there are found only Planer's Lamprey, Gudgeon, Tench, Roach, Dace, Chub, Minnow, Loach, Ten-spined Stickleback, Pike, Perch, Bull-head, and perhaps

Rudd; in Devonshire and Cornwall the Chub is absent, Roach and Rudd occur only in Slapton Ley and a few other localities in Devonshire, and Tench are said not to be indigenous; in Cornwall the Perch is local and introduced. Quite a number of species are absent from Wales west of the Severn system, and in addition to the species already mentioned as confined to eastward streams of England the Gudgeon, the Bleak, and the Ruffe are notable absentees from the Lake District.

The fact that species which have a very wide and a very similar distribution on the mainland of Europe and Asia, have very dissimilar distributions in the British Islands, can only be explained by supposing that our islands were connected with each other and with continental Europe comparatively recently, when our eastern, and probably our southern, streams were tributaries of continental rivers and received from them the fishes which they contained; only nine or ten of these had reached Ireland before it became a separate island, and the distribution of the rest in Britain at varying rates according to circumstances has not yet proceeded long enough to spread them all over the island.

The British Isles stand upon a plateau, the edge whereof coincides roughly with the 100-fathom line. From this the bottom of the sea slopes rather suddenly down to depths of 1000 to 1500 fathoms, ancient river-beds can be traced in the Irish and English Channels, which prove conclusively that at one time the streams of northern France and southern England were tributaries of one " English Channel " River, and that those of the south and east of Ireland joined the same main " Irish Channel " River as the Severn and

other streams of the West of England and Wales.
Moreover, these submerged river channels, towards
the edge of the plateau, get deeper and deeper and
can be traced down nearly to its base, so that it is
clear that we have here the ancient coast-line of
Western Europe, which must have been bounded by
steep cliffs, perhaps seven or eight thousand feet high,
and as our islands are connected with Iceland and
Greenland by a submarine ridge of less than half that
depth those countries also probably formed at that
time part of the European continent.   There is reason
to believe that the conditions just described may have
existed throughout a considerable part of the Tertiary
Epoch, and that at times during this period Greenland
may have been connected with America, thus uniting
the two continents; these were also connected *via*
Asia, and there is good evidence of at least one
remote and one comparatively recent union of Asia
and North America over Bering's Straits.   Thus the
general similarity of the freshwater fishes of the
Palæarctic and Nearctic regions is not to be wondered
at.

During the greater part of this time the climate
was warmer than it is at the present day, but then
came the Glacial Epoch.   It would be out of place
here to enter into the causes of this climatic change,
but it seems clear that at a comparatively recent date
the whole of Northern Europe, including our islands
except the south of England, became covered with
ice; then a submergence of the land perhaps brought
the sea right to the edge of the ice-sheets, so that any
true freshwater fishes which may have been in our
rivers would have perished.   At the end of this cold
period, a gradual elevation of the land took place,

FIG. 26.—Restoration of Pleistocene Geography of the British Isles, showing the Coast-line coincident with the contour of 80 fathoms (*after Jukes-Browne*).

culminating in the union of our islands with each other and with the Continent, and then a subsidence followed which gave to our islands approximately their present outline.

Speculations as to the exact time occupied by these changes are futile, but it is probable that the whole duration of the Tertiary Period should be expressed in tens of millions of years and that of the Glacial Epoch in tens, or at most hundreds, of thousands. Possibly about 100,000 years may have elapsed since the end of the Glacial Epoch, and our final separation from the Continent was of still more recent date.

As soon as the ice-sheet had begun to disappear Char must have commenced running up into the lakes which were formed, and as the elevation of the land proceeded new lakes appeared and in turn became inhabited by Char. All this time the climate was gradually getting warmer, and the southern limit of anadromous Char was receding northward; then the Sea-trout reached our islands, and these in turn began to form fluviatile colonies in our lakes and rivers.

At the present day the southern limit of anadromous Char, namely, Iceland and the northern coasts of Europe, nearly coincides with the northern limit of anadromous Trout. Thus we can understand why some of our lakes contain Char, but no Trout. Troutless lakes are fairly plentiful in the Highlands of Scotland, although their number has greatly decreased in recent years through stocking with Trout. In Ireland, Lough Gortyglass in Clare is an example of a lake which contains Char, but had no Trout until they were introduced about thirty years ago by Messrs. W.

and F. B. Henn.   The latter gentleman tells me that there is a fall almost immediately below the lake in the stream which flows out of it and enters the estuary of the Shannon, and that this fall is too steep for Trout to ascend, although they are plentiful enough below it.   It seems probable that Char entered Lough Gortyglass at a time when arctic conditions were prevalent and the lake was almost on the sea-level; the subsequent elevation which brought lakes such as Corrib and Currane into existence may have formed the falls which barred the entrance of Trout to Lough Gortyglass when the commencement of the amelioration of the arctic climate brought Sea-trout to the shores of Clare.   The lake is now situated at an altitude of from 200 to 250 feet above the level of the sea, a less elevation than one would anticipate for a troutless lake so far south, if movements of elevation and subsidence had been nearly uniform over the British Isles since the Glacial Epoch began to pass away.   Indeed, the absence of Trout from Gortyglass, as well as the presence of Char in lakes scarcely above the sea-level in Galway and Kerry, leads me to believe that the subsidence that succeeded the elevation that joined Britain to Ireland and to the Continent has been greater in the south and west of Ireland than in any other part of the British Isles. The Whitefish probably established themselves at about the same time as the Char, before the arctic conditions had passed away.

When the elevation had reached its maximum and our islands once more formed part of the Continent, the coast-line may have approximated to the 100-fathom contour   Our eastern rivers were tributaries of the Rhine, which flowed northward along what is now the bed of the North Sea ; we may suppose that they received from the Rhine all the fishes which then inhabited it and which were able to ascend them ; but that the rivers of Eastern England should have received so many fishes and those of Scotland so few is not a little curious Of course, in those days the rivers of Scotland had not cut their way down to their present level and were more rapid, but this must also have applied more or less to those of Yorkshire, and one cannot help feeling that if sluggish fish like Bream and Rudd were able to establish themselves in the latter, the lively and active Dace ought to have got into the former.   I rather incline to the theory that for this last brief union with the Continent the 40-fathom rather than the 100-fathom line may represent the coast in the North Sea area, and this seems to me in harmony with the geological evidence that elevation and gain of land has proceeded in the north of Britain since it became an island.

In the south-west the 100-fathom line may well represent the former coast-line, and there can be little doubt that our southern rivers received fishes from the northern rivers of France *via* the " English Channel River." This fish-fauna was evidently a poorer one than that of the Rhine, but probably most of the true freshwater fishes of the Stour and Avon came to them from this source, and the comparative poverty of the fish-fauna of Devon and Cornwall is due to the unsuitability of their small and rapid streams for most of the species, although here, in contrast to Scotland, the Dace is indigenous.

A glance at the map will show that Ireland probably got its fishes *via* the Irish Channel River from the Severn, or other Western rivers, or it may have obtained part of its fauna by a temporary confluence of the rivers of the Irish and English Channels. It may here be remarked that the submerged river-beds do not necessarily represent the courses of the rivers during the short period of our last union with the Continent.

If the routes by which freshwater fishes reached our islands have been indicated, their further distribution has been accomplished chiefly by those geological changes which lead to the capture of the tributaries of one river by another, or by the agency of man, who at one time greatly appreciated many of the species as food, and has indeed spread the Carp, a native of Eastern Asia, over the greater part of our islands, and must be held in some degree responsible for the present range of the Tench, Pike, and Perch. The Minnow and Loach, probably the only really indigenous freshwater fishes in Scotland north of Loch Lomond and the Firth of Forth, are small species which thrive in little brooks, and are therefore the more likely to spread rapidly, being transferred from one system to another by slight changes in the head-waters which would not affect the inhabitants of the lower parts of the rivers.

Lest it be thought I am too ready to assume such changes, I may say that geologists seem to regard it as established that the Severn System has enormously aggrandized itself by the capture of tributaries of the Thames, Trent, and other rivers, and it is quite probable that the Irish Channel River got its fishes by the cutting back of one or more of its English tributaries, thus capturing from the Rhine System.

Water-spouts, transference of spawn by birds, and other accidental methods have never, in my opinion, succeeded in establishing a species in a river system from which it was previously absent.

PLATE XXXVII

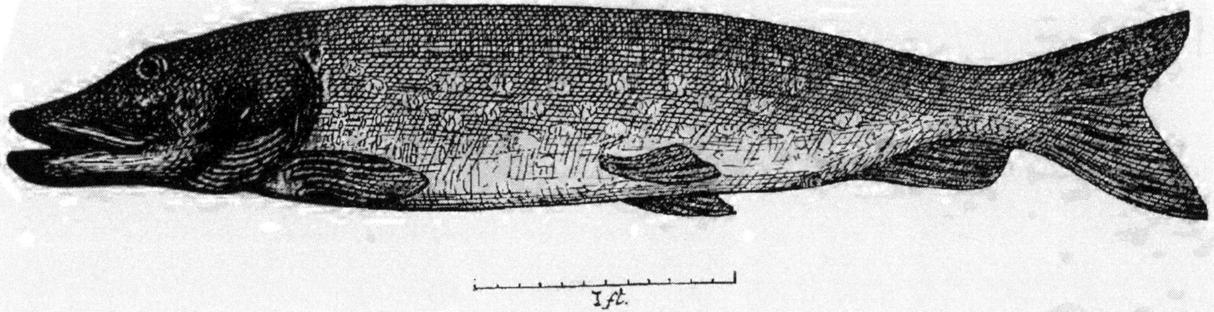

1 ft.

THE GREAT PIKE OF WHITTLESEA MERE

APPENDIX

THE GREAT PIKE OF WHITTLESEA MERE

THE above is the title of a chromolithograph which was published by Messrs. Hanhart, and which has been sent to me by Mr. R. B. Marston, Editor of the *Fishing Gazette*, who received it from a correspondent a few years ago. The picture bears the inscription, "Caught in Whittlesea Mere, Weight 52 Pounds. I. M." A scale is annexed, from which it appears that the total length of the fish, from the end of the snout to the fork of the caudal fin, is 52 inches, 13 of which are occupied by the head.

This very large Pike was taken when Whittlesea Mere was drained in 1851; the figure on the opposite page is copied from the picture above-mentioned, and will give some idea of its proportions. The weight stated, 52 lbs., is certainly not excessive if the fish is drawn correctly to scale, for a Pike of this length, in good condition, might well weigh 60 lbs. or more; but this fish has rather the appearance of one out of condition, or which had seen its best days and was getting lanky and large-headed.

This may be regarded as the record English Pike, and it is interesting to note that although Whittlesea Mere is in Cambridgeshire it is on the borders of Lincolnshire, and that Walton wrote that, in England, Lincolnshire boasted to have the biggest Pikes.

INDEX

Ingram Content Group UK Ltd.
Milton Keynes UK
UKHW051136050423
419466UK00026B/129